Delmar's Test Preparation Series

Collision Repair/Refinish

Non-Structural Analysis and Damage Repair (Test B3)

Technical Advisor
Boyce H. Dwiggins

Delmar Publishers

RECEIVED

FEB 0 3 1999

Linn State Technical College
Library/TLC

an International Thomson Publishing company I(T)P

Albany • Bonn • Boston • Cincinnati • Detroit • London • Madrid
Melbourne • Mexico City • New York • Pacific Grove • Paris • San Francisco
Singapore • Tokyo • Toronto • Washington

TL
152
.M573
1999

NOTICE TO THE READER

Publisher does not warrant or guarantee any of the products described herein or perform any independent analysis in connection with any of the product information contained herein. Publisher does not assume, and expressly disclaims, any obligation to obtain and include information other than that provided to it by the manufacturer.

The reader is expressly warned to consider and adopt all safety precautions that might be indicated by the activities described herein and to avoid all potential hazards. By following the instructions contained herein, the reader willingly assumes all risks in connections with such instructions.

The publisher makes no representation or warranties of any kind, including but not limited to, the warranties of fitness for particular purpose or merchantability, nor are any such representations implied with respect to the material set forth herein, and the publisher takes no responsibility with respect to such material. The publisher shall not be liable for any special, consequential, or exemplary damages resulting, in whole or part, from the readers' use of, or reliance upon, this material.

Delmar Staff
Publisher: Alar Elken
Acquisitions Editor: Vern Anthony
Editorial Assistant: Betsy Hough
Marketing Manager: Mona Caron

COPYRIGHT © 1998
By Delmar Publishers
an International Thomson Publishing Company

The ITP logo is a trademark under license

Printed in the United States of America

For more information, contact:

Delmar Publishers
3 Columbia Circle, Box 15015
Albany, New York 12212-5015

International Thomson Publishing Europe
Berkshire House 168-173
High Holborn
London, WC1V7AA
England

Thomas Nelson Australia
102 Dodds Street
South Melbourne, 3205
Victoria, Australia

Nelson Canada
1120 Birchmont Road
Scarborough, Ontario
Canada M1K 5G4

Online Services

Delmar Online
To access a wide variety of Delmar products and services on the World Wide Web, point your browser to:
http://www.delmar.com
or email: info@delmar.com

thomson.com
To access International Thomson Publishing's home site for information on more than 34 publishers and 20,000 products, point your browser to:
http://www.thomson.com
or email: findit@kiosk.thomson.com

A service of I(T)P®

International Thomson Editores
Campos Eliseos 385, Piso 7
Col Polanco
11560 Mexico D F Mexico

International Thomson Publishing GmbH
Königswinterer Strasse 418
53227 Bonn
Germany

International Thomson Publishing Asia
221 Henderson Road #05-10
Henderson Building
Singapore 0315

International Thomson Publishing–Japan
Hirakawacho Kyowa Building, 3F
2-2-1 Hirakawacho
Chiyoda-ku, 102 Tokyo
Japan

All rights reserved. No part of this work covered by the copyright hereon may be reproduced or used in any form or by any means—graphic, electronic, or mechanical, including photocopying, recording, taping, or information storage and retrieval systems—without the written permission of the publisher.

1 2 3 4 5 6 7 8 9 10 XXX 02 01 00 99 98

ISBN 0-7668-0568-9

Contents

Preface vi

Section 1 The History of ASE

History 1
    NIASE 1
    The Series and Individual Tests. 2
    A Brief Chronology. 2
    By the Numbers 3
    ASE 4

Section 2 Take and Pass Every ASE Test

ASE Testing. 7
    Who Writes the Questions? 7
    Testing. 8
    Be Test-Wise 8
    Before the Test. 8
    Objective Tests 9
    Taking an Objective Test. 9
    During the Test 10
    Review Your Answers 10
    Don't Be Distracted. 10
    Use Your Time Wisely. 11
    Don't Cheat 11
    Be Confident. 11
    Anxiety and Fear. 12
    Getting Rid of Fear 12
    Effective Study 13
    Make Study Definite 13
    The Urge to Learn. 14
    Concentrate 14
        Get Sufficient Sleep 15
        Arrange Your Area 15
        Don't Daydream 15
    Study Regularly. 16

Keep a Record 17
Scoring the ASE Test 17
Understand the Test Results 18
   Non-Structural Analysis and Damage Repair (Test B3) 18
"Average" 19
So, How Did You Do? 19

Section 3 Are You Sure You're Ready for B3 Test?

Pretest 21
   Answers to the Test Questions 23
   Explanations for Selected Test Answers 24
Types of Questions 25
   Multiple-Choice Questions 25
   EXCEPT Questions 25
   Technician A, Technician B Questions 26
   Questions with a Figure 27
   Most Likely Questions 28
   LEAST Likely Questions 28
   Summary 28
Testing Time Length 29
   Monitor Your Progress 29
   Registration 29

Section 4 An Overview of the System

Non-Structural Analysis and Damage Repair (Test B3) 31
   Task List 31
      A. Preparation (5 Questions) 31
      B. Outer Body Panel Repairs, Replacements, and
         Adjustments (16 Questions) 32
      C. Metal Finishing and Body Filling (6 Questions) 33
      D. Movable Glass and Hardware (4 Questions) 33
      E. Welding and Cutting (10 Questions) 33
      F. Plastic Repair (9 Questions) 34
   Overview 35

Section 5 Sample Test for Practice

Sample Test 59

Section 6 Additional Test Questions for Practice

Additional Test Questions 87

Section 7 Appendices

Answers to the Test Questions for the Sample Test
Section 5. 101
Answers to the Test Questions for the Additional Test Questions
Section 6. 102
Glossary 103

Preface

This book is just one of a comprehensive series designed to prepare technicians to take and pass every ASE test. Delmar's series covers all of the Automotive tests A1 through A8 as well as Advanced Engine Performance L2 and Parts Specialist P2; the series covers the five Collision Repair tests and the eight Medium/Heavy Duty truck tests.

Each book in this series has the same features designed for the technician to use and succeed with. Before a word was written, we met with and surveyed technicians and shop owners who have taken ASE tests and have used other preparatory materials. We found that they wanted, first and foremost, *lots* of practice tests and questions. Each book in our series contains a general knowledge pretest, a sample test, and additional practice questions. You will be hard-pressed to find a test prep book with more questions for you to practice with. And, we have worked hard to ensure that these questions match the ASE style in types of questions, quantities, and level of difficulty.

Technicians also told us that they wanted to understand the ASE test and to have practical information about what they should expect. We have provided that as well, including a history of ASE and a section devoted to helping the technician "Take and Pass Every ASE Test" with case studies, test-taking strategies, and test formats.

Finally, techs wanted refresher information and reference. Each of our books includes an overview section that is referenced to the task list. The complete task lists for each test appear in each book for the user's reference. There is also a complete glossary of terms for each booklet.

So whether you're looking for a sample test and a few extra questions to practice with or a complete introduction to ASE testing and support for preparing thoroughly, this book series is an excellent answer.

We hope you benefit from this book and that you pass every ASE test you take!

Your comments, both positive and negative, are most certainly encouraged! Please contact us at:
Automotive Editor
Delmar Publishers
3 Columbia Circle
Box 15015
Albany, NY 12212-5015

The History of ASE

History

Originally known as The National Institute for Automotive Service Excellence (NIASE), today's ASE was founded in 1972 as a non-profit, independent entity dedicated to improving the quality of automotive service and repair through the voluntary testing and certification of automotive technicians. Until that time, consumers had no way of distinguishing between competent and incompetent automotive mechanics. In the mid-1960s and early 1970s, efforts were made by several automotive industry affiliated associations to respond to this need. Though the associations were non-profit, many regarded certification test fees merely as a means of raising additional operating capital. Also some, having a vested interest, resulted in test scores heavily weighted in the favor of its members.

NIASE

From these efforts a new independent, non-profit association, the National Institute for Automotive Service Excellence (NIASE) was established much to the credit of two educators, George R. Kinsler, Director of Program Development for the Wisconsin Board of Vocational and Adult Education in Madison, WI, and Myron H. Appel, Division Chairman at Cypress College in Cypress, CA.

Early efforts were to encourage voluntary certification in eight areas:

TEST	TITLE
I.	Engines, Block Assembly, Cooling and Lube Systems, Induction, Ignition, & Exhaust
II.	Automatic Transmissions
III.	Manual Transmissions, Drive Line & Rear Axles
IV.	Steering, Suspension, & Wheels
V.	Brakes
VI.	Body/Chassis, Electrical Systems
VII.	Heating and Air Conditioning
VIII.	Engine Tune-Up

In early NIASE tests, Mechanic A, Mechanic B type questions were used. Over the years the trend has not changed, but in mid-1984 the term was changed to Technician A, Technician B to better emphasize sophistication of the skills needed to perform successfully in the modern motor vehicle industry. In certain tests, the term used is Estimator A/B, Painter A/B, or Parts Specialist A/B. At about that same time, the logo was changed from "The Gear" to "The Blue Seal," and the organization adopted the acronym ASE for Automotive Service Excellence.

Since those early beginnings, several other related trades have been added. ASE now administers a comprehensive series of certification exams for automotive and light truck repair technicians, medium and heavy truck repair technicians, alternate fuels technicians, engine machinists, collision repair technicians, school bus repair technicians, and parts specialists.

The Series and Individual Tests

- Automotive and Light Truck Technician; consisting of: Engine Repair - Automatic Transmission/Transaxle - Manual Drive Train and Axles - Suspension and Steering - Brakes - Electrical/Electronic Systems - Heating and Air Conditioning - Engine Performance
- Medium and Heavy Truck Technician; consisting of: Gasoline Engines - Diesel Engines - Drive Train - Brakes - Suspension and Steering - Electrical/Electronic Systems - HVAC - Preventive Maintenance Inspection
- Alternate Fuels Technician; consisting of: Compressed Natural Gas—Light Vehicles
- Advanced Series; consisting of: Automobile Advanced Engine Performance and Advanced Diesel Engine Electronic Diesel Engine Specialty
- Collision Repair Technician; consisting of: Painting and Refinishing - Non-Structural Analysis and Damage Repair - Structural Analysis and Damage Repair - Mechanical and Electrical Components - Damage Analysis and Estimating
- Engine Machinist Technician; consisting of: Cylinder Head Specialist - Cylinder Block Specialist - Assembly Specialist
- School Bus Repair Technician; consisting of: Body Systems and Special Equipment - Drive Train - Brakes - Suspension and Steering - Electrical/Electronic Systems - Heating and Air Conditioning
- Parts Specialist; consisting of: Automobile Parts Specialist - Medium/Heavy Truck Parts Specialist

A Brief Chronology

1970–1971   Original questions were prepared by a group of forty auto mechanics teachers from public secondary schools, technical institutes, community colleges, and private vocational schools. These questions were then professionally edited by testing specialists at Educational Testing Service (ETS) at Princeton, New Jersey, and thoroughly reviewed by training specialists associated with domestic and import automotive companies.

1971   July: About eight hundred mechanics tried out the original test questions at experimental test administrations.

1972   November and December: Initial NIASE tests administered. There were 163 test centers. The original automotive test series consisted of four tests containing eighty questions each. Three hours were allotted for each test. Those who passed all four tests were designated Certified General Auto Mechanic (GAM).

1973   April and May: Test 4 was increased to 120 questions. Time was extended to four hours for this test. There are now 182 test centers. Shoulder patch insignias were made available.

The History of ASE

	November: Automotive series expanded to five tests. Heavy Duty Truck series of six tests introduced.
1974	November: Automatic Transmission (Light Repair) test modified. Name changed to Automatic Transmission.
1975	May: Collision Repair series of two tests is introduced.
1978	May: Automotive recertification testing is introduced.
1979	May: Heavy Duty Truck recertification testing is introduced.
1980	May: Collision Repair recertification testing is introduced.
1982	May: Test administration providers switched from Educational Testing Service (ETS) to American College Testing (ACT). Name of Automobile Engine Tune-Up test changed to Engine Performance test.
1984	May: New logo was introduced. ASE's "The Blue Seal" replaced NIASE's "The Gear." All reference to Mechanic A, Mechanic B was changed to Technician A, Technician B.
1990	November: The first of the Engine Machinist test series was introduced.
1991	May: The second test of the Engine Machinist test series was introduced. November: The third and final Engine Machinist test was introduced.
1992	May: Name of Heavy-Duty Truck Test series changed to Medium/Heavy Truck Test series.
1993	May: Automotive Parts Specialist test introduced. Collision Repair expanded to six tests. Light Vehicle Compressed Natural Gas test introduced. Limited testing begins in English-speaking provinces of Canada.
1994	May: Advanced Engine Performance Specialist test introduced.
1996	May: First three tests for School Bus Technician Test series introduced. November: A Collision Repair test is added.
1997	May: A Medium/Heavy Truck test is added.
1998	May: A diesel advanced engine test is introduced: Electronic Diesel Engine Diagnosis Specialist. A test is added to the School Bus Test series.

By the Numbers

Following are the approximate number of ASE technicians currently certified by category. The numbers may vary from time to time but are reasonably accurate for any given period. More accurate data may be obtained from ASE, which provides updates twice each year, in May and November after the Spring and Fall test series.

There are more than 338,000 Automotive Technicians with over 87,000 at Master Technician (MA) status. There are 47,000 Truck Technicians with over 19,000 at Master Technician (MT) status. There are 46,000 Collision Repair/Refinish Technicians with 7,300 at Master Technician (MB) status. There are 1,200 Estimators. There are 6,700 Engine Machinists with over 2,800 at Master Machinist Technician (MM) status. There are also 28,500 Automobile Advanced Engine Performance Technicians and over 2,700 School Bus Technicians for a combined total of more than 403,000 Repair Technicians. To this number, add over 22,000 Automobile Parts Specialists, and over 2,000 Truck Parts Specialists for a combined total of over 24,000 parts specialists.

There are over 6,400 ASE Technicians holding both Master Automotive Technician and Master Truck Technician status, of which 350 also hold Master Body Repair status. Almost 200 of these Master Technicians also hold Master Machinist status and five Technicians are certified in all ASE specialty areas.

Almost half of ASE certified technicians work in new vehicle dealerships (45.3%). The next greatest number work in independent garages with 19.8%. Next is tire dealerships with 9%, service stations at 6.3%, fleet shops at 5.7%, franchised volume retailers at 5.4%, paint and body shops at 4.3%, and specialty shops at 3.9%.

Of over 400,000 automotive technicians on ASE's certification rosters, almost 2,000 are female. The number of female technicians is increasing at a rate of about 20% each year. Women's increasing interest in automotive mechanics is further evidenced by the fact that, according to the National Automobile Dealers Association (NADA), they influence 80% of the decisions of the purchase of a new automobile and represent 50% of all new car purchasers. Also, it is interesting to note that 65% of all repair and maintenance service customers are female.

The typical ASE certified technician is 36.5 years of age, is computer literate, deciphers a half-million pages of technical manuals, spends one hundred hours per year in training, holds four ASE certificates, and spends about $27,000 for tools and equipment. Twenty-seven percent of today's skilled ASE certified technicians attended college, many having earned an Associate of Science degree in Automotive Technology.

ASE

ASE's mission is to improve the quality of vehicle repair and service in the United States through the testing and certification of automotive repair technicians. Prospective candidates register for and take one or more of ASE's thirty-three exams. The tests are grouped into specialties for automobile, medium/heavy truck, school bus, and collision repair technicians as well as engine machinists, alternate fuels technicians, and parts specialists.

Upon passing at least one exam and providing proof of two years of related work experience, the technician becomes ASE certified. A technician who passes a series of exams earns ASE Master Technician status. An automobile technician, for example, must pass eight exams for this recognition.

The tests, conducted twice a year at over seven hundred locations around the country, are administered by American College Testing (ACT). They stress real-world diagnostic and repair problems. Though a good knowledge of theory is helpful to the technician in answering many of the questions, there are no questions specifically on theory. Certification is valid for five years. To retain certification, the technician must be retested to renew his or her certificate.

The automotive consumer benefits because ASE certification is a valuable yardstick by which to measure the knowledge and skills of individual technicians, as well as their commitment to their chosen profession. It is also a tribute to the repair facility employing ASE certified technicians. ASE certified technicians are permitted to wear blue and white ASE shoulder insignia, referred to as the "Blue Seal of Excellence," and carry credentials listing their areas of expertise. Often employers display their technicians' credentials in the customer waiting area. Customers look for facilities that display ASE's Blue Seal of Excellence logo on outdoor signs, in the customer waiting area, in the telephone book (Yellow Pages), and in newspaper advertisements.

The tests stress repair knowledge and skill. All test takers are issued a score report. In order to earn ASE certification, a technician must pass one or more of the exams and present proof of two years of relevant hands-on work experience. ASE certifications are valid for five years, after which time technicians must retest in order to keep up with changing technology and to remain in the ASE program. A nominal registration and test fee is charged.

The History of ASE

To become part of the team that wear ASE's Blue Seal of Excellence®, please contact:

National Institute for Automotive Service Excellence
13505 Dulles Technology Drive
Herndon, VA 20171-3421

2 Take and Pass Every ASE Test

ASE Testing

Participating in an Automotive Service Excellence (ASE) voluntary certification program gives you a chance to show your customers that you have the "know-how" needed to work on today's modern vehicles. The ASE certification tests allow you to compare your skills and knowledge to the automotive service industry's standards for each specialty area.

If you are the "average" automotive technician taking this test, you are in your mid-thirties and have not attended school for about fifteen years. That means you probably have not taken a test in many years. Some of you, on the other hand, have attended college or taken postsecondary education courses and may be more familiar with taking tests and with test-taking strategies. There is, however, a difference in the ASE test you are preparing to take and the educational tests you may be accustomed to.

Who Writes the Questions?

The questions on an educational test are generally written, administered, and graded by an educator who may have little or no practical hands-on experience in the test area. The questions on all ASE tests are written by service industry experts familiar with all aspects of the subject area. ASE questions are entirely job-related and designed to test the skills that you need to know on the job.

The questions originate in an ASE "item-writing" workshop where service representatives from domestic and import automobile manufacturers, parts and equipment manufacturers, and vocational educators meet in a workshop setting to share their ideas and translate them into test questions. Each test question written by these experts is reviewed by all of the members of the group. The questions deal with the practical problems of diagnosis and repair that are experienced by technicians in their day-to-day hands-on work experiences.

All of the questions are pretested and quality-checked in a non-scoring section of tests by a national sample of certifying technicians. The questions that meet ASE's high standards of accuracy and quality are then included in the scoring sections of future tests. Those questions that do not pass ASE's stringent test are sent back to the workshop or are discarded. ASE's tests are monitored by an independent proctor and are administered and machine-scored by an independent provider, American College Testing (ACT).

Testing

If you think about it, we are actually tested on about everything we do. As infants, we were tested to see when we could turn over and crawl, later when we could walk or talk. As adolescents, we were tested to determine how well we learned the material presented in school and in how we demonstrated our accomplishments on the athletic field. As working adults, we are tested by our supervisors on how well we have completed an assignment or project. As nonworking adults, we are tested by our family on everyday activities, such as housekeeping or preparing a meal. Testing, then, is one of those facts of life that begins in the cradle and follows us to the grave.

Testing is an important fact of life that helps us to determine how well we have learned our trade. Also, tests often help us to determine what opportunities will be available to us in the future. To become ASE certified, we are required to take a test in every subject in which we wish to be recognized.

Be Test-Wise

In spite of the widespread use of tests, most technicians are not very test-wise. An ability to take tests and score well is a skill that must be acquired. Without this knowledge, the most intelligent and prepared technician may not do well on a test.

We will discuss some of the basic procedures necessary to follow in order to become a test-wise technician. Assume, if you will, that you have done the necessary study and preparation to score well on the ASE test.

Different approaches should be used for taking different types of tests. The different basic types of tests include: essay, objective, multiple-choice, fill in the blank, true-false, problem solving, and open-book. All ASE tests are of the four-part multiple-choice type.

Before discussing the multiple-choice type test questions, however, there are a few basic principles that should be followed before taking any test.

Before the Test

Do not arrive late. Always arrive well before your test is scheduled to begin. Allow ample time for the unexpected, such as traffic problems, so you will arrive on time and avoid the unnecessary anxiety of being late.

Always be certain to have plenty of supplies with you. For an ASE test, three or four sharpened soft lead (#2) pencils, a pocket pencil sharpener, erasers, and a watch are all that are required.

Do not listen to pretest chatter. When you arrive early, you may hear other technicians testing each other on various topics or making their best guess as to the probable test questions. At this time, it is too late to add to your knowledge. Also the rhetoric may only confuse you. If you find it bothersome, take a walk outside the test room to relax and loosen up.

Read and listen to all instructions. It is important to read and listen to the instructions. Make certain that you know what is expected of you. Listen carefully to verbal instructions and pay particular attention to any written instructions on the test paper. Do not dive into answering questions only to find out that you have answered the wrong question by not following instructions carefully. It is difficult to make a high score on a test if you answer the wrong questions.

These basic principles have been violated in most every test ever given. Try to remember them. They are essential for success.

Objective Tests

A test is called an objective test if the same standards and conditions apply to everyone taking the test and there is only one "right" answer to each question. Objective tests primarily measure your ability to recall information. A well-designed objective test can also test your ability to understand, analyze, interpret, and apply your knowledge. Objective tests include true-false, multiple-choice, fill in the blank, and matching questions.

Objective questions, not generally encountered in a classroom setting, are frequently used in standardized examinations. Objective tests are easy to grade and also reduce the amount of paperwork necessary to administer. The objective tests are used in entry-level programs or when very large numbers are being tested. ASE's tests consist exclusively of four-part multiple-choice objective questions in all of their tests. Here are some of the basic principles of taking an objective test.

Taking an Objective Test

The principles of taking an objective test are somewhat different from those used in other types of tests. You should first quickly look over the test to determine the number of questions, but do not try to read through all of the questions. In an ASE test, there are usually between forty and eighty questions, depending on the subject matter. Read through each question before marking your answer. Answer the questions in the order they appear on the test. Leave the questions blank that you are not sure of and move on to the next question. You can return to those unanswered questions after you have finished the others. They may be easier to answer at a later time after your mind has had additional time to consider them on a subconscious level. In addition, you might find information in other questions that will help you to answer some of them.

Do not be obsessed by the apparent pattern of responses. For example, do not be influenced by a pattern like **d, c, b, a, d, c, b, a** on an ASE test.

There is also a lot of folk wisdom about taking objective tests. For example, there are those who would advise you to avoid response options that use certain words such as *all, none, always, never, must,* and *only,* to name a few. This, they claim, is because nothing in life is exclusive. They would advise you to choose response options that use words that allow for some exception, such as *sometimes, frequently, rarely, often, usually, seldom,* and *normally.* They would also advise you to avoid the first and last option (A and D) because test writers, they feel, are more comfortable if they put the correct answer in the middle (B and C) of the choices. Another recommendation often offered is to select the option that is either shorter or longer than the other three choices because it is more likely to be correct. Some would advise you to never change an answer since your first intuition is usually correct.

Although there may be a grain of truth in this folk wisdom, ASE test writers try to avoid them and so should you. There are just as many **A** answers as there are **B** answers, just as many **D** answers as **C** answers. As a matter of fact, ASE tries to balance the answers at about 25 percent per choice **A, B, C,** and **D**. There is no intention to use "tricky" words, such as outlined above. Put no credence in the opposing words "sometimes" and "never," for example. When used in an ASE type question, one or both may be correct; one or both may be incorrect.

There are some special principles to observe on multiple-choice tests. These tests are sometimes challenging because there are often several choices that may seem possible, and it may be difficult to decide on the correct choice. The best strategy, in this case, is to first determine the correct answer before looking at the options. If you see the answer you decided on, you should still examine the options to make sure that none seem more correct than yours. If you do not know or are not sure of the answer, read each option very carefully and try to eliminate those options that you know to be wrong. That way, you can often arrive at the correct choice through a process of elimination

If you have gone through all of the test and you still do not know the answer to some of the questions, then guess. Yes, guess. You then have at least a 25 percent chance of being correct. If you leave the question blank, you have no chance. In ASE tests, there is no penalty for being wrong. As the late President Franklin D. Roosevelt once advised a group of students, "It is common sense to take a method and try it. If it fails, admit it frankly and try another. But above all, try something."

During the Test

Mark your bubble sheet clearly and accurately. One of the biggest problems an adult faces in test-taking, it seems, is in placing their answer in the correct spot on a bubble sheet. Make certain that you mark in your answer for, say, question 21, in the space on the bubble sheet designated for the answer for question 21. A correct response in the wrong bubble will probably be incorrect. Remember, the answer sheet is machine-scored and can only "read" what you have bubbled in. Also, do not bubble in two answers for the same question. For example, if you feel the answer to a particular question is A but think it may be C, do not bubble in both choices. Even if either A or C is correct, a double answer will score as an incorrect answer. It's better to take a chance with your best guess.

Review Your Answers

If you finish answering all of the questions on a test ahead of time, go back and review the answers of those questions that you were not sure of. You can often catch careless errors by using the remaining time to review your answers.

Don't Be Distracted

At practically every test, some technicians will invariably finish ahead of time and turn their papers in long before the final call. Do not let them distract or intimidate you. Either they knew too little and could not finish the test, or they were very self-confident and thought they knew it all. Perhaps they were trying to impress the proctor or other technicians about how much they know. Often you may hear them later talking about the information they knew all the while but forgot to respond on their answer sheet.

Use Your Time Wisely

It is not wise to use less than the total amount of time that you are allotted for a test. If there are any doubts, take the time for review. Any product can usually be made better with some additional effort. A test is no exception. It is not necessary to turn in your test paper until you are told to do so.

Don't Cheat

Some technicians may try to use a "crib sheet" during a test. Others may attempt to read answers from another technician's paper. If you do that, you are unquestionably assuming that someone else has a correct answer. You probably know as much, maybe more, than anyone else in the test room. Trust yourself. If you're still not convinced, think of the consequences of being caught. Cheating is foolish. You may get caught. If you are caught, you have failed the test.

Be Confident

The first and foremost principle in taking a test is that you need to know what you are doing, to be test-wise. It will now be presumed that you are a test-wise technician and are now ready for some of the more obscure aspects of test-taking.

An ASE style test requires that you use the information and knowledge at your command to solve a problem. This generally requires a combination of information similar to the way you approach problems in the real world. Most problems, it seems, typically do not fall into neat textbook cases. New problems are often difficult to handle, whether they are encountered inside or outside the test room.

An ASE test also requires that you apply methods taught in class as well as those learned on the job to solve problems. These methods are akin to a well-equipped tool box in the hands of a skilled technician. You have to know what tools to use in a particular situation, and you must also know how to use them. In an ASE test, you will need to be able to demonstrate that you are familiar with and know how to use the tools.

You should begin a test with a completely open mind. At times, however, you may have to move out of your normal way of thinking and be creative to arrive at a correct answer. If you have diligently studied for at least one week prior to the test, you have bombarded your mind with a wide assortment of information. Your mind will be working with this information on a subconscious level, exploring the interrelationships among various facts, principles, and ideas. This prior preparation should put you in a creative mood for the test.

In order to reach your full potential, you should begin a test with the proper mental attitude and a high degree of self-confidence. You should think of a test as an opportunity to document how much you know about the various tasks in your chosen profession. If you have been diligently studying the subject matter, you will be able to take your test in serenity because your mind will be well organized. If you are confident, you are more likely to do well because you have the proper mental attitude. If, on the other hand, your confidence is low, you are bound to do poorly. It is a self-fulfilling prophecy.

Perhaps you have heard athletic coaches talk about the importance of confidence when competing in sports. Mental confidence helps an athlete to perform at the highest level and gain an advantage over competitors. Taking a test is much like an athletic event. You are competing against yourself, in a certain sense, because you will be trying to approach perfection in determining your answers. As in any competition, you should aim your sights high and be confident that you can reach the apex.

Anxiety and Fear

Many technicians experience anxiety and fear at the very thought of taking a test. Many worry, become nervous, and even become ill at test time because of the fear of failure. Many often worry about the criticism and ridicule that may come from their employer, relatives, and peers. Some worry about taking a test because they feel that the stakes are very high. Those who spent a great amount of time studying may feel they must get a high grade to justify their efforts. The thought of not doing well can result in unnecessary worry. They become so worried, in fact, that their reasoning and thinking ability is impaired, actually bringing about the problem they wanted to avoid.

The fear of failure should not be confused with the desire for success. It is natural to become "psyched-up" for a test in contemplation of what is to come. A little emotion can provide a healthy flow of adrenaline to peak your senses and hone your mental ability. This improves your performance on the test and is a very different reaction from fear.

Most technician's fears and insecurities experienced before a test are due to a lack of self-confidence. Those who have not scored well on previous tests or have no confidence in their preparation are those most likely to fail. Be confident that you will do well on your test and your fears should vanish. You will know that you have done everything possible to realize your potential.

Getting Rid of Fear

If you have previously experienced fear of taking a test, it may be difficult to change your attitude immediately. It may be easier to cope with fear if you have a better understanding of what the test is about. A test is merely an assessment of how much the technician knows about a particular task area. Tests, then, are much less threatening when thought of in this manner. This does not mean, however, that you should lower your self-esteem simply because you performed poorly on a test.

You can consider the test essentially as a learning device, providing you with valuable information to evaluate your performance and knowledge. Recognize that no one is perfect. All humans make mistakes. The idea, then, is to make mistakes before the test, learn from them, and avoid repeating them on the test. Fortunately, this is not as difficult as it seems. Practical questions in this study guide include the correct answers to consider if you have made mistakes on the practice test. You should learn where you went wrong so you will not repeat them in the ASE test. If you learn from your mistakes, the stage is set for future growth.

If you understood everything presented up until now, you have the knowledge to become a test-wise technician, but more is required. To be a test-wise technician, you not only have to practice these principles, you have to diligently study in your task area.

Effective Study

The fundamental and vital requirement to induce effective study is a genuine and intense desire to achieve. This is more basic than any rule or technique that will be given here. The key requirement, then, is a driving motivation to learn and to achieve.

If you wish to study effectively, first develop a desire to master your studies and sincerely believe that you will master them. Everything else is secondary to such a desire.

First, build up definite ambitions and ideals toward which your studies can lead. Picture the satisfaction of success. The attitude of the technician may be transformed from merely getting by to an earnest and energetic effort. The best direct stimulus to change may involve nothing more than the deliberate planning of your time. Plan time to study.

Another drive that creates positive study is an interest in the subject studied. As an automotive technician, you can develop an interest in studying particular subjects if you follow these four rules:

1. Acquire information from a variety of sources. The greater your interest in a subject, the easier it is to learn about it. Visit your local library and seek books on the subject you are studying. When you find something new or of interest, make inexpensive photocopies for future study.
2. Merge new information with your previous knowledge. Discover the relationship of new facts to old known facts. Modern developments in automotive technology take on new interest when they are seen in relation to present knowledge.
3. Make new information personal. Relate the new information to matters that are of concern to you. The information you are now reading, for example, has interest to you as you think about how it can help.
4. Use your new knowledge. Raise questions about the points made by the book. Try to anticipate what the next steps and conclusions will be. Discuss this new knowledge, particularly the difficult and questionable points, with your peers.

You will find that when you study with eager interest, you will discover it is no longer work. It is pleasure and you will be fascinated in what you study. Studying can be like reading a novel or seeing a movie that overcomes distractions and requires no effort or willpower. You will discover that the positive relationship between interest and efforts works both ways. Even though you perhaps begin your studies with little or no interest, simply staying with it helped you to develop an interest in your studies.

Obviously, certain subject matter studies are bound to be of little or no interest, particularly in the beginning. Parts of certain studies may continue to be uninteresting. An honest effort to master those subjects, however, nearly always brings about some level of interest. If you appreciate the necessity and reward of effective studying, you will rarely be disappointed. Here are a few important hints for gaining the determination that is essential to carrying good conclusions into actual practice.

Make Study Definite

Decide what is to be studied and when it is to be studied. If the unit is discouragingly long, break it into two or more parts. Determine exactly what is involved in the first part and learn that. Only then should you proceed to the next part. Stick to a schedule.

The Urge to Learn

Make clear to yourself the relation of your present knowledge to your study materials. Determine the relevance with regard to your long-range goals and ambitions.

Turn your attention away from real or imagined difficulties as well as other things that you would rather be doing. Some major distractions are thoughts of other duties and of disturbing problems. These distractions can usually be put aside, simply shunted off by listing them in a notebook. Most technicians have found that by writing interfering thoughts down, their minds are freed from annoying tensions.

Adopt the most reasonable solution you can find or seek objective help from someone else for personal problems. Personal problems and worry are often causes of ineffective study. Sometimes there are no satisfactory solutions. Some manage to avoid the problems or to meet them without great worry. For those who may wish to find better ways of meeting their personal problems, the following suggestions are offered:

1. Determine as objectively and as definitely as possible where the problem lies. What changes are needed to remove the problem, and which changes, if any, can be made? Sometimes it is wiser to alter your goals than external conditions. If there is no perfect solution, explore the others. Some solutions may be better than others.
2. Seek an understanding confidant who may be able to help analyze and meet your problems. Very often talking over your problems with someone in whom you have confidence and trust will help you to arrive at a solution.
3. Do not betray yourself by trying to evade the problem or by pretending that it has been solved. If social problem distractions prevent you from studying or doing satisfactory work, it is better to admit this to yourself. You can then decide what can be done about it.

Once you are free of interferences and irritations, it is much easier to stay focused on your studies.

Concentrate

To study effectively, you must concentrate. Your ability to concentrate is governed, to a great extent, by your surroundings as well as your physical condition. When absorbed in study, you must be oblivious to everything else around you. As you learn to concentrate and study, you must also learn to overcome all distractions. There are three kinds of distractions you may face:

1. Distractions in the surrounding area, such as motion, noise, and the glare of lights. The sun shining through a window on your study area, for example, can be very distracting.

    Some technicians find that, for effective study, it is necessary to eliminate visual distractions as well as noises. Others find that they are able to tolerate moderate levels of auditory or visual distractions.

    Make sure your study area is properly lighted and ventilated. The lighting should be adequate but should not shine directly into your eyes or be visible out of the corner of your eye. Also, try to avoid a reflection of the lighting on the pages of your book.

    Whether heated or cooled, the environment should be at a comfortable level. For most, this means a temperature of 78°F–80°F (25.6°C–26.7°C) with a relative humidity of 45 to 50 percent.

2. Distractions arising from your body, such as a headache, fatigue, and hunger. Be in good physical condition. Eat wholesome meals at regular times. Try to eat with your family or friends whenever possible. Meal time should be your recreational period. Do not eat a heavy meal for lunch and do not resume studies immediately after eating lunch. Just after lunch, try to get some regular exercise, relaxation, and recreation. A little exercise on a regular basis is much more valuable than a lot of exercise only on occasion.
3. Distractions of irrelevant ideas, such as how to repair the garden gate when you are studying for an automotive-related test.

The problems associated with study are no small matter. These problems of distractions are generally best dealt with by a process of elimination. A few important rules for eliminating distractions follow.

Get Sufficient Sleep

You must get plenty of rest even if it means dropping certain outside activities. Avoid cutting in on your sleep time; you will be rewarded in the long run. If you experience difficulty going to sleep, do something to take your mind off your work and try to relax before going to bed. Some suggestions that may help include a little reading, a warm bath, a short walk, a conversation with a friend, or writing that overdue letter to a distant relative. If sleeplessness is an ongoing problem, consult a physician. Do not try any of the sleep remedies on the market, particularly if you are on medication, without approval of your physician.

If you still have difficulty studying, a final rule may help. Sit down in a favorable place for studying, open your books, and take out your pencil and paper. In a word, go through the motions.

Arrange Your Area

Arrange your chair and work area. To avoid strain and fatigue, whenever possible, shift your position occasionally. Try to be comfortable; however, avoid being too comfortable. It is near impossible to study rigorously when settled back in a large easy chair or reclining leisurely on a sofa.

When studying, it is essential to have a plan of action, a time to work, a time to study, and a time for pleasure. If you schedule your day and adhere to the schedule, you will eliminate most of your efforts and worries. A plan that is followed, then, soon becomes the easy and natural routine of the day. Most technicians find it useful to have a definite place and time to study. A particular table and chair are always used for study and intellectual work. This place will, then, come to mean study. To be seated in that particular location at a regular scheduled time will automatically lead you to assume a readiness for study.

Don't Daydream

Daydreaming or mind-wandering is an enemy of effective study. Daydreaming is frequently due to an inadequate understanding of words. Use the Glossary or a dictionary to look up the troublesome word. Another frequent cause of daydreaming is a deficient background in the present subject matter. When this is the problem, go back and review the subject matter to obtain the necessary foundation. Just one hour of concentrated study is equivalent to ten hours with frequent lapses of daydreaming. Be on guard against mind-wandering, and pull yourself back into focus on every occasion.

Study Regularly

A system of regularity in study is believed by many scholars to be the secret of success. The daily time schedule must, however, be determined on an individual basis. You must decide how many hours each day you can devote to your studies. Few technicians really are aware of where their leisure time is spent. An accurate account of how your days are presently being spent is an important first step toward creating an effective daily schedule.

	\multicolumn{7}{c}{WEEKLY SCHEDULE}						
	SUN	MON	TUES	WED	THU	FRI	SAT
6:00							
6:30							
7:00							
7:30							
8:00							
8:30							
9:00							
9:30							
10:00							
10:30							
11:00							
11:30							
NOON							
12:30							
1:00							
1:30							
2:00							
2:30							
3:00							
3:30							
4:00							
4:30							
5:00							
5:30							
6:00							
6:30							
7:00							
7:30							
8:00							
8:30							
9:00							
9:30							
10:00							
10:30							
11:00							
11:30							

The convenient form is for keeping an hourly record of your week's activities. If you fill in the schedule each evening before bedtime, you will soon gain some interesting and useful facts about yourself and your use of your time. If you think over the causes of wasted time, you can determine how you might better spend your time. A practical schedule can be set up by using the following steps:

Mark your fixed commitments, such as work, on your schedule. Be sure to include classes and clubs. Do you have sufficient time left? You can arrive at an estimate of the time you need for studying by counting the hours used during the present week. An often used formula, if you are taking classes, is to multiply the number of hours you spend in class by two. This provides time for class studies. This is then added to your work hours. Do not forget time allocation for travel.

Fill in your schedule for meals and studying. Use as much time as you have available during the normal workday hours. Do not plan, for example, to do all of your studying between 11:00 pm and 1:00 am. Try to select a time for study that you can use every day without interruption. You may have to use two or perhaps three different study periods during the day.

List the things you need to do within a time period. A one-week time frame seems to work well for most technicians. The question you may ask yourself is: "What do I need to do to be able to walk into the test next week, or next month, prepared to pass?"

Break down each task into smaller tasks. The amount of time given to each area must also be settled. In what order will you tackle your schedule? It is best to plan the approximate time for your assignments and the order in which you will do them. In this way, you can avoid the difficulties of not knowing what to do first and of worrying about the other things you should be doing.

List your tasks in the empty spaces on your schedule. Keep some free time unscheduled so you can deal with any unexpected events, such as a dental appointment. You will then have a tentative schedule for the following week. It should be flexible enough to allow some units to be rearranged if necessary. Your schedule should allow time off from your studies. Some use the promise of a planned recreational period as a reward for motivating faithfulness to a schedule. You will more likely lose control of your schedule if it is packed too tightly.

Keep a Record

Keep a record of what you actually do. Use the knowledge you gain by keeping a record of what you are actually doing so you can create or modify a schedule for the following week. Be sure to give yourself credit for movement toward your goals and objectives. If you find that you can not study productively at a particular hour, modify your schedule so as to correct that problem.

Scoring the ASE Test

You can gain a better perspective about tests if you know and understand how they are scored. ASE's tests are scored by American College Testing (ACT), a non-partial, non-biased organization having no vested interest in ASE or in the automotive industry. Each question carries the same weight as any other question. For example, if there are fifty questions, each is worth 2 percent of the total score. The passing grade is 70 percent. That means you must correctly answer thirty-five of the fifty questions to pass the test.

Understand the Test Results

The test results can tell you:
- Where your knowledge equals or exceeds that needed for competent performance, or
- Where you might need more preparation.

The test results *cannot* tell you:
- How you compare with other technicians, or
- How many questions you answered correctly.

Your ASE test score report will show your correct answers as percentage ranges. These ranges provide information about your performance in each area of the test. However, because there may be a different number of questions in each area of the test, a high percentage in an area with few questions may not offset a low percentage in an area with many questions.

It may be noted that one does not "fail" an ASE test. The technician that does not pass is simply told "More Preparation Needed." Though large differences in percentages may indicate problem areas, it is important to consider how many questions were asked in each area. Since each test evaluates all phases of the work involved in a service specialty, you should be prepared in each area. A low score in one area could keep you from passing an entire test.

Note that a typical test will contain the number of questions indicated above each content area's description. For example:

Non-Structural Analysis and Damage Repair (Test B3)

Content Area	Questions	Percent of Test
A. Preparation	5	10%
B. Outer Body Panel Repairs, Replacements, and Adjustments	16	32%
C. Metal Finishing and Body Filling	6	12%
D. Movable Glass and Hardware	4	8%
E. Welding and Cutting	10	20%
F. Plastic Repair	9	18%
Total	50	100%

"Average"

There is no such thing as average. You can not determine your overall test score by adding the percentages given for each task area and dividing by the number of areas. It doesn't work that way because there generally are not the same number of questions in each task area. A task area with, say, 20 questions, for example, counts more toward your total score than a task area with 10 questions.

So, How Did You Do?

Your test report should give you a good picture of your results and a better understanding of your task areas of strength and weakness.

If you fail to pass the test, you may take it again at any time it is scheduled to be administered. You are the only one who will receive your test score. Test scores will not be given over the telephone by ASE nor will they be released to anyone without your written permission.

3 Are You Sure You're Ready for B3 Test?

Pretest

The purpose of this pretest is to determine the amount of review that you may require prior to taking the ASE Collision Repair/Refinish: Non-Structural Analysis and Damage Repair (Test B3). If you answer all of the pretest questions correctly, complete the sample test for practice along with the additional test questions.

If two or more of your answers to the pretest questions are incorrect, study section 4: An Overview of the System before continuing with the sample test for practice and additional test questions.

The pretest answers and explanations are located at the end of the pretest.

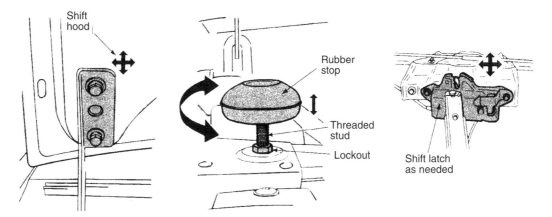

1. When adjusting the hood alignment, as shown above, you should make adjustments at the hood:
   A. latch.
   B. striker.
   C. hinges.
   D. bumpers.

2. When removing door trim panels, you should use a(n):
   A. pneumatic knife.
   B. electric knife.
   C. forked tool.
   D. slide hammer.

21

3. When visually positioning a non-structural outer body panel, technician A makes sure that the gaps between the panels are tapered.
   Technician B makes sure that the gaps between the panels are even.
   Who is right?
   A. A only
   B. B only
   C. Both A and B
   D. Neither A nor B

4. After recovering the refrigerant from an A/C system, technician A uses plastic wrap with wire ties to keep moisture out of the system components.
   Technician B replaces the receiver-drier if the system has been open to the atmosphere for several days.
   Who is right?
   A. A only
   B. B only
   C. Both A and B
   D. Neither A nor B

5. All of the following methods are used to strip paint from metal surfaces of a vehicle EXCEPT:
   A. sanding.
   B. blasting.
   C. chemical stripping.
   D. hot tanking.

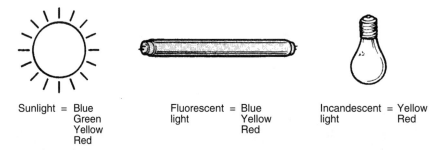

Sunlight = Blue Green Yellow Red
Fluorescent light = Blue Yellow Red
Incandescent light = Yellow Red

6. What type of light, as shown above, should be used when evaluating the color of a finish?
   A. Sunlight
   B. Incandescent light
   C. Fluorescent light
   D. Drop light

7. Technician A says to try color sanding and polishing if acid spotting is not too severe.
   Technician B says that sanding and refinishing might be needed.
   Who is right?
   A. A only
   B. B only
   C. Both A and B
   D. Neither A nor B

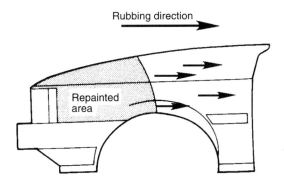

8. Technician A says rubbing compound should be used to smooth surface imperfections on the vehicle panel above.
   Technician B says polishing compound should be used to bring out a gloss.
   Who is right?
   A. A only
   B. B only
   C. Both A and B
   D. Neither A nor B

9. The LEAST likely cause of paint blushing is:
   A. excessive air pressure.
   B. using too fast of a thinner.
   C. painting in hot, humid conditions.
   D. improper surface cleaning.

10. The LEAST likely cause of poor flow when painting is:
    A. improper air pressure.
    B. a dirty paint gun.
    C. improper paint mixture.
    D. improper gun stroke.

11. Improper paint adhesion can be caused by all of the following EXCEPT:
    A. improper surface preparation.
    B. poor adhesion of old paint.
    C. uneven body work.
    D. rust under paint film.

12. Dents can be filled with all of the following EXCEPT:
    A. aluminum fillers.
    B. spot and glazing putties.
    C. plastic fillers.
    D. fiberglass fillers.

Answers to the Test Questions

1. C, 2. C, 3. B, 4. B, 5. D, 6. A, 7. C, 8. C, 9. D, 10. C, 11. C, 12. B

Explanations for Selected Test Answers

Question #1
Answer A is wrong. The hood latch mechanism keeps the hood closed and releases the hood when activated. Adjusting it will not align the hood to the fenders or cowl.
Answer B is wrong. The hood striker engages the hood latch when closing the hood. Adjusting it will not align the hood to the fenders or cowl.
Answer C is correct. The hood hinge adjustment controls the general position of the hood. By loosening the hood hinge bolts, you can move the hood right, left, front, or back.
Answer D is wrong. The hood bumpers are used to set hood height at the front of the hood. Adjusting them will not align the hood to the fenders or cowl.

Question #3
The right way to position a body panel is to align the panel so the gap between the panels is even. Technician A is wrong, and technician B is right.
Answer A is wrong. Technician A is wrong.
Answer B is correct. Technician B is right.
Answer C is wrong. Only technician B is right.
Answer D is wrong. Technician B is right.

Question #5
Answer A is wrong. Machine sanding or grinding is suitable for removing old finish from small areas.
Answer B is wrong. Blasting uses air powered tools for forcing sand, plastic beads, or another material onto the surface for paint removal.
Answer C is wrong. A chemical paint remover can be used for stripping large areas of paint.
Answer D is correct. Hot tanking is commonly used to rid small parts of dirt and grease, not to remove paint from body panels.

Question #7
Answer A is wrong. Technician A is right. Spotting can sometimes be removed with wet sanding and compounding if there is only minor damage.
Answer B is wrong. Technician B is right. Sanding and refinishing may be needed if contamination has reached the metal or substrate.
Answer C is correct. Both technicians are right.
Answer D is wrong. Both technicians are right.

Question #9
Answer A is wrong. Excessive air pressure when painting will cause blushing.
Answer B is wrong. Using too fast of a thinner when mixing paints will cause blushing.
Answer C is wrong. Painting in hot, humid conditions will cause moisture to condense on the wet paint film and cause blushing.
Answer D is correct. Improper surface cleaning can cause many paint problems, but it will not cause blushing.

Question #11
Answer A is wrong. Improper surface preparation will cause poor adhesion of the paint.
Answer B is wrong. Poor adhesion of old paint will cause the new paint to also have poor adhesion.
Answer C is correct. Uneven body work will not cause poor adhesion of the paint, only imperfect body lines.
Answer D is wrong. Rust under paint will cause very poor adhesion.

Types of Questions

ASE certification tests are often thought of as being tricky. They may seem to be tricky if you do not completely understand what is being asked. The following examples will help you recognize certain types of ASE questions and avoid common errors.

Each test is made up of forty to eighty multiple-choice questions. Multiple-choice questions are an efficient way to test knowledge. To answer them correctly, you must think about each choice as a possibility, and then choose the one that best answers the question. To do this, read each word of the question carefully. Do not assume you know what the question is about until you have finished reading it.

Multiple-Choice Questions

One type of multiple choice question has three wrong answers and one correct answer. The wrong answers, however, may be almost correct, so be careful not to jump at the first answer that seems to be correct. If all the answers seem to be correct, choose the answer that is the most correct. If you readily know the answer, this kind of question does not present a problem. If you are not sure of the answer, analyze the question and the answers. For example:

Question 1:
   A rocker panel is a structural member of which vehicle construction type?
   A. Front-wheel drive
   B. Pickup truck
   C. Space-frame
   D. Full-frame
   Analysis:
This question asks for a specific answer. By carefully reading the question, you will find that it asks for a construction type that uses the rocker panel as a structural part of the vehicle.
Answer A is wrong. Front-wheel drive is not a vehicle construction type.
Answer B is wrong. A pickup truck is not a type of vehicle construction.
Answer C is correct. Space frame is a unibody design creating structural integrity by welding parts together, such as the rocker panels, but does not require exterior cosmetic panels installed for full strength.
Answer D is wrong. Full frame describes a body-over-frame construction type that relies on the frame assembly for structural integrity.
   Therefore, the correct answer is C. If the question was read quickly and the words "construction type" were passed over, answer A may have been selected.

EXCEPT Questions

Another type of question used on ASE tests has answers that are all correct except one. The correct answer for this type of question is the answer that is wrong. The word "EXCEPT" will always be in capital letters. You must identify which of the choices is the wrong answer. If you read quickly through the question, you may overlook what the question is asking and answer the question with the first correct statement. This will make your answer wrong. EXCEPT questions are generally found at the end of the test, but be on the alert for them throughout the test. An example of this type of question and the analysis is as follows:

Question 2:

All of the following are a tool for the analysis of structural damage EXCEPT:

A. height gauge.
B. tape measure.
C. dial indicator.
D. tram gauge.

Analysis:

The question really requires you to identify the tool that is not used for analyzing structural damage. All tools given in the choices are used for analyzing structural damage except one. This question presents two basic problems for the test-taker who reads through the question too quickly. It may be possible to read over the word "EXCEPT" in the question or not think about which type of damage analysis would use answer C. In either case, the correct answer may not be selected. To correctly answer this question, you should know what tools are used for the analysis of structural damage. If you cannot immediately recognize the incorrect tool, you should be able to identify it by analyzing the other choices.

Answer A is wrong. A height gauge *may* be used to analyze structural damage.

Answer B is wrong. A tape measure *is* a tool that may be used to analyze structural damage.

Answer C is correct. A dial indicator may be used as a damage analysis tool for moving parts, such as wheels, wheel hubs, and axle shafts, but would *not* be used to measure structural damage.

Answer D is wrong. A tram gauge *is* used to measure structural damage.

Technician A, Technician B Questions

The type of question that is most popularly associated with an ASE test is the "Technician A says... Technician B says... Who is right?" type. In this type of question, you must identify the correct statement or statements. To answer this type of question correctly, you must carefully read each technician's statement and judge it on its own merit to determine if the statement is true.

Typically, this type of question begins with a statement about some analysis or repair procedure. This is followed by two statements about the cause of the problem, proper inspection, identification, or repair choices. You are asked whether the first statement, the second statement, both statements, or neither statement is correct. Analyzing this type of question is a little easier than the other types because there are only two ideas to consider although there are still four choices for an answer. An example of this type of question and the analysis of it is as follows:

Question 3:

Structural dimensions are being measured.
Technician A says comparing measurements from one side to the other is enough to determine the damage.
Technician B says a tram gauge can be used when a tape measure cannot measure in a straight line from point to point.
Who is right?

A. A only
B. B only
C. Both A and B
D. Neither A nor B

Analysis:
With some vehicles built asymmetrical, side-to-side measurements are not always equal. The manufacturer's specifications need to be verified with a dimension chart before reaching any conclusions about the structural damage.
Answer A is wrong. Technician A's statement is wrong. A tram gauge would provide a point-to-point measurement when a part, such as a strut tower or air cleaner, interrupts a direct line between the points.
Answer B is correct. Technician B is correct. A tram gauge can be used when a tape measure cannot be used to measure in a straight line from point to point.
Answer C is wrong. Since Technician A is not correct, C cannot be the correct answer.
Answer D is wrong. Since Technician B is correct, D cannot be the correct answer.

Questions with a Figure

About 10 percent of ASE questions will have a figure, as shown in the following example:
Question 4:

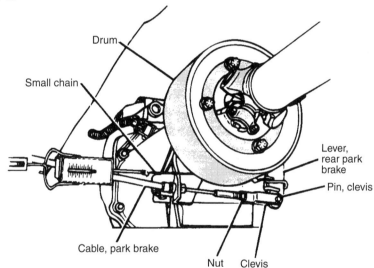

In the measurement shown in the figure above:
A. the driveshaft center support bearing wear is measured.
B. the propeller shaft parking brake is adjusted.
C. the brake shoes must be adjusted before this measurement is taken.
D. the parking brake is released while this measurement is taken.

Analysis:
Answer A is wrong. The driveshaft center support bearing wear is not being measured in this figure.
Answer B is correct. The propeller shaft parking brake adjustment is being performed.
Answer C is wrong. The brake shoes need not be adjusted before adjusting the propeller shaft parking brake.
Answer D is wrong. The parking brake is applied during the propeller shaft parking brake adjustment procedures.

Most Likely Questions

Most likely questions are somewhat difficult in that only one choice is correct while the other three choices are nearly correct. An example of a most likely cause question is as follows:

Question 5:

The most likely cause of reduced turbocharger boost pressure may be a:
A. westgate valve stuck closed.
B. westgate valve stuck open.
C. leaking westgate diaphragm.
D. disconnected westgate linkage.

Analysis:

Answer A is wrong. A westgate valve stuck closed increases turbocharger boost pressure.
Answer B is correct. A westgate valve stuck open decreases turbocharger boost pressure.
Answer C is wrong. A leaking westgate valve diaphragm increases turbocharger boost pressure.
Answer D is wrong. A disconnected westgate valve linkage will increase turbocharger boost pressure.

LEAST Likely Questions

Notice that in most likely questions there is no capitalization. This is not so with LEAST likely type questions. For this type of question, look for the choice that would be least likely cause the described situation. Read the entire question carefully before choosing your answer. LEAST likely questions are generally found at the end of the test, but be on the alert for them throughout the test. An example is as follows:

Question 6:

What is the LEAST likely cause of a bent push rod?
A. Excessive engine speed
B. A sticking valve
C. Excessive valve guide clearance
D. A worn rocker arm stud

Analysis:

Answer A is wrong. Excessive engine speed may cause a bent push rod.
Answer B is wrong. A sticking valve may cause a bent push rod.
Answer C is correct. Excessive valve clearance will not generally cause a bent push rod.
Answer D is wrong. A worn rocker arm stud may cause a bent push rod.

Summary

There are no four-part multiple-choice ASE questions having "none of the above" or "all of the above" choices. ASE does not use other types of questions, such as fill-in-the-blank, completion, true-false, word-matching, or essay. ASE does not require you to draw diagrams or sketches. If a formula or chart is required to answer a question, it is provided for you. There are no ASE questions that require you to use a pocket calculator.

Testing Time Length

An ASE test session is four hours and fifteen minutes. You may attempt from one to a maximum of four tests in one session. It is recommended, however, that no more than a total of 225 questions be attempted at any test session. This will allow for just over one minute for each question.

Visitors are not permitted at any time. If you wish to leave the test room, for any reason, you must first ask permission. If you finish your test early and wish to leave, you are permitted to do so only during specified dismissal periods.

Monitor Your Progress

You should monitor your progress and set an arbitrary limit to how much time you will need for each question. This should be based on the number of questions you are attempting. It is suggested that you wear a watch because some facilities may not have a clock visible to all areas of the room.

Registration

Test centers are assigned on a first-come, first-served basis. To register for an ASE certification test, it is suggested that you enroll at least six weeks before the scheduled test date. This should provide sufficient time to ensure you a spot in the test center. It should also give you enough time for study in preparation for the test. Test sessions are offered by ASE twice each year, in May and November, at over six hundred sites across the United States.

To register, contact Automotive Service Excellence/American College Testing at:

ASE/ACT

P.O. Box 4007

Iowa City, IA 52243

4 An Overview of the System

Non-Structural Analysis and Damage Repair (Test B3)

The following section includes the task areas and task lists for this test and a written overview of the topics covered in the test.

The task list describes the actual work you should be able to do as a technician that you will be tested on by the ASE. This is your key to the test and you should review this section carefully. We have based our sample test and additional questions upon these tasks, and the overview section will also support your understanding of the task list. ASE advises that the questions on the test may not equal the number of tasks listed; the task lists tell you what ASE expects you to know how to do and be ready to be tested upon.

Task List

A. Preparation (5 Questions)

Task 1 Review damage report; analyze damage to determine appropriate methods for overall repair.

Task 2 Lift, raise, and position vehicle to perform repairs.

Task 3 Remove outside trim and moldings as necessary; store reusable parts.

Task 4 Remove damaged or undamaged inside trim and moldings as necessary; store reusable parts.

Task 5 Remove undamaged nonstructural body panels and components that may interfere with or be damaged during repair.

Task 6 Remove all vehicle mechanical and electrical components that may interfere with or be damaged during repair.

Task 7 Protect panels and parts adjacent to repair areas to prevent damage during repair.

Task 8 Remove dirt, grease, wax, and decals from areas to be repaired.

Task 9 Remove corrosion protection, undercoatings, sealers, and other protective coatings as necessary to perform repairs.

Task 10 Remove repairable plastics and other parts that are recommended for off-vehicle repair.

Task 11 Identify potential safety and environmental concerns associated with vehicle components and systems; e.g., ABS, air bags (SRS), refrigerants, coolants, etc.

B. Outer Body Panel Repairs, Replacements, and Adjustments (16 Questions)

Task 1 Determine the extent of the direct and indirect damage and the direction of impact; plan the methods and order of repair.

Task 2 Remove and replace bolted, bonded, and welded panels or panel assemblies.

Task 3 Determine the extent of damage to aluminum body panels; repair, weld, or replace in accordance with manufacturer's specifications.

Task 4 Remove, replace, and align hood, hood hinges, and hood latch/lock.

Task 5 Remove, replace, and align deck lid, lid hinges, and lid latch/lock.

Task 6 Remove and replace doors, tailgates, hatches, liftgates, latch/lock assemblies, and hinges.

Task 7 Remove, replace, and align bumpers, reinforcements, guards, isolators, and mounting hardware.

Task 8 Check and adjust clearances of front fenders, header, and other panels.

Task 9 Check door hinge condition; check door frames for proper fit; check and adjust door clearances.

Task 10 Rough-out contours of damaged panel to a surface condition suitable for metal finishing and body filling.

Task 11 Weld cracked or torn steel body panels; reweld broken welds; replace molding studs.

Task 12 Apply protective coatings and sealants to restore corrosion protection.

Task 13 Remove damaged sections of steel body panels; weld in replacements in accordance with vehicle manufacturer's specifications; inspect intrusion beams.

Task 14 Repair or replace door skins in accordance with vehicle manufacturer's specifications; inspect intrusion beams.

Task 15 Replace or repair plastic panels in accordance with vehicle manufacturer's/industry specifications.

Task 16 Restore sealers, mastic, sound deadeners, and foam fillers.

Task 17 Diagnose and repair water leaks, dust leaks, wind noise, squeaks, and rattles.

Task 18 Install interior and exterior trim and moldings.

C. Metal Finishing and Body Filling (6 Questions)

Task 1 Grind off paint from the damaged area of a body panel.

Task 2 Pick and file the damaged area of a body panel to eliminate surface irregularities.

Task 3 Prepare surface for application of body filler material.

Task 4 Heat shrink stretched panel areas to proper contour.

Task 5 Cold shrink stretched panel areas to proper contour.

Task 6 Mix plastic filler.

Task 7 Apply plastic body filler; rough shape during curing.

Task 8 Sand cured plastic filler to contour.

D. Movable Glass and Hardware (4 Questions)

Task 1 Inspect, adjust, repair, or replace window regulators, run channels, glass (including electrically heated glass), power mechanisms, and related controls.

Task 2 Repair or replace electrically driven power sunroofs and related controls.

Task 3 Inspect, repair or replace, and adjust removable manually operated roof panels and hardware.

Task 4 Diagnose and repair water leaks, dust leaks, and noises; inspect, repair, and replace weather stripping.

Task 5 Inspect, repair, and install convertible top and related mechanisms.

E. Welding and Cutting (10 Questions)

Task 1 Identify weldable and nonweldable materials used in vehicle construction.

Task 2   Understand the limitations of welding and cutting high-strength steels and other metals.

Task 3   Determine the correct welding process (GMAW [MIG] compression/resistance spot, oxyacetylene, GTAW [TIG]), electrode, wire type, diameter, and gas to be used in specific welding situations.

Task 4   "Tune" the MIG welder by adjusting for proper electrode stickout, voltage, polarity, flow rate, and wire-feed speed required for material being welded.

Task 5   Identify safety considerations: eye protection, proper clothing, shock hazards, fumes, MSDS, etc. before beginning any welding operation.

Task 6   Apply knowledge of the proper procedures for safely handling gas cylinders.

Task 7   Insure proper work clamp (ground) location.

Task 8   Use proper gun-to-joint angle and direction of gun travel for welds being made in all positions.

Task 9   Protect vehicle components including computers and other electronic modules from possible welding and cutting damage.

Task 10   Clean the metal to be welded; assure good metal fit-up; apply weld-through primer.

Task 11   Perform the right weld joint type (butt, lap, etc.) for the weld being made.

Task 12   Determine the right type of weld (continuous, stitch/pulse, tack, plug, spot, etc.) for each specific welding operation.

Task 13   Identify the causes of spits and sputters, burn-through, lack of penetration, cracks in metal, porosity, incomplete fusion, excessive spatter, distortion, waviness of bead, and failure of wire feed; make necessary adjustments.

Task 14   Identify proper cutting process for different materials and locations in accordance with manufacturer's recommendations.

F. Plastic Repair (9 Questions)

Task 1   Identify the types of plastic(s); determine repairability.

Task 2   Identify the proper plastic repair procedure; clean and prepare the surface of plastic parts in accordance with manufacturer's/industry guidelines.

Task 3   Repair rigid plastic parts by welding.

Task 4   Repair rigid plastic parts with urethane or epoxy adhesives, with and without reinforcement materials.

Task 5   Repair flexible plastic parts by welding.

Task 6   Repair flexible plastic parts with urethane or epoxy adhesives, with and without reinforcement materials.

Task 7   Repair holes and cuts in rigid and flexible plastic parts using backing materials and adhesives.

Task 8   Retexture plastic parts.

Task 9   Repair vinyl-clad urethane foam parts.

Task 10 Reshape and shrink flexible exterior plastic parts.

Task 11 Remove damaged areas from rigid exterior SMC (sheet molded compound) panels; repair with partial panel installation.

Task 12 Repair deep gouges, holes, and cracks in SMC (sheet molded compound) panels.

Task 13 Replace bonded non-SMC (sheet molded compound) type body panels; straighten or align panel supports.

Task 14 Replace bonded non-SMC plastic body panels; straighten or align panel supports.

Task 15 Prepare repaired areas for refinishing.

Overview

A. When starting work, refer to the estimate to get guidance on how to begin. The estimator will have determined which parts need to be repaired and which should be replaced. An estimate, also called a damage report or appraisal, calculates the cost of parts, materials, and labor for repairing a collision damaged vehicle. You would use this information and shop manuals to remove and replace parts efficiently. The estimate is an important reference tool for doing repairs. It must be followed. The insurance company and estimators have both determined which parts must be repaired. If you fail to follow the estimate, the insurance company may not pay for your work.

Another important function of the estimate is that it serves as a basis for writing the work order or operational plan. It is usually prepared from the written estimate of the estimator and a visual inspection by the shop supervisor. The work order outlines the procedure that should be taken to put the vehicle back in preaccident condition.

The estimate is used to order new parts. You might want to make sure all ordered parts have arrived. Compare new parts on hand with the parts list. If anything is missing, have the parts person order them. This will save time and prevent your work from being tied up while waiting for parts.

B. Some important pieces of shop equipment are jacks and hoists used to raise the vehicle in the air for easier working conditions. The different types of jacks are the mechanical scissor jack, the hydraulic bottle jack, and the hydraulic service jack, all of which are used for lifting the front, rear, and side of a vehicle. There are also many different types of hoists; there are drive-on hoists, twin-post hoists, side-post hoists, and the old style center-post hoists. All of these can help the technician by putting the vehicles at a comfortable working height. A hoist is a specialized piece of equipment, and a technician needs to know exactly how to use it, especially if there are any vehicle weight problems. It is important to always use the mechanical safeties. After lifting a vehicle to the desired height, always lower the hoist onto the mechanical safeties.

It is important to note that the quality of today's hoists, as well as the number of safety devices on each model, makes them very safe to operate. There are specific contact points where the weight of the vehicle can be supported. The right lifting points can be found in the vehicle's service manual.

C. Before beginning disassembly, any exterior trim that is in the way or that might get damaged while sanding or performing other body work should be removed. In most cases, moldings or nameplates are secured with bolts, screws, retaining clips, welded-on nuts, or adhesives. Take care when removing trim to keep it from getting bent or damaged. Any trim that can be reused should be stored for later use. Some trim may require special tools for removal; sometimes a molding tool and heat gun are used to help remove moldings and nameplates.

D. Before beginning disassembly, any inside trim that is in the way or that might get damaged while performing work should be removed. Various pieces are used in the passenger compartment for appearance and safety. Most are held in by snap-in clips, push pins, or small screws. Sometimes screw heads are covered by small plastic plugs. Screws can also be hidden under protruding parts. Your service manual will give locations for the fasteners holding interior trim parts. Any trim that can be reused should be stored for later use.

E. Vehicles with major damage must often have their frame or body structures straightened. Vehicle straightening involves using high power hydraulic or pneumatic equipment, mechanical clamps, and chains to bring the frame or body structure back into its original shape. As a general rule, only remove the parts that prevent you from getting to the area of the vehicle being repaired. Depending on the construction of the vehicle and the location and degree of the damage, there will be cases where it will be more convenient to remove bolt-on parts before proceeding with the repair. Carefully analyze the vehicle and damage to determine what must be removed. It is sometimes best to remove bolt-on parts before putting the vehicle on the rack. You might have better access to the fasteners.

F. As a general rule, only remove the parts that prevent you from getting to the area of the vehicle being repaired. At one time, you often had to remove the suspension and driveline completely from a unibody vehicle before putting it on the straightening machine. With most of the current straightening systems and such accessories as the engine holder, this is no longer necessary. Major straightening operations can be done with major mechanical parts intact.

Depending on the construction of the vehicle and the location and degree of the damage, there will be cases where it will be more convenient to remove bolt-on parts before proceeding with the repair. Carefully analyze the vehicle and damage to determine what must be removed. Take time to carefully study the locations of engine and transmission mounts, suspension mounts, and whether or not these parts themselves are damaged.

G. There are no clear ground rules on when to mask and when to remove parts. This would include parts like trim, moldings, and door handles. The decision to remove or mask depends upon the design of the vehicle and the expectations of the customer. If a part can be removed easily, it is better to remove it than try to mask around it. Also, if you cannot sand and clean right up to it, the part should be removed. If it will be difficult to mask the part, you might save time by removing it from the vehicle. Masking can be accomplished with tape or a spray-on masking coating that washes off with soap and water. Each part will require an individual decision.

H. The vehicle should be washed to remove any mud, dirt, or other water-soluble contaminants before brought in the shop. Hose down the vehicle, sponge it with detergent and water, and then rinse thoroughly. Wash the top, front, deck, and sides; allow the vehicle to dry.

Before the job is sanded, use a specially blended wax and grease remover or recommended solvent to thoroughly clean the surface. Be sure to thoroughly clean all areas where a heavy wax buildup can be a problem, such as around trim, moldings, door handles, and the radio antenna, and behind the bumpers. Paint will not adhere properly to a waxy surface. Wax and silicone can penetrate beneath the surface. This contamination is not easily detectable. It is wise to assume that it is present; therefore, always include some wax and grease cleaner or detergent in the sanding water.

Special attention is required for tar, gasoline, battery acid, antifreeze, and brake fluid stains. These can also penetrate well beneath the surface of old paint films, and their residues must be removed during the sanding operation.

I. It is often necessary to remove the paint film, undercoat, sealer, or other coatings covering body panel joints to find the location of the spot welds. To do this, remove the paint using a dual action (DA) sander with medium grit paper or use a scuff wheel in a grinder. A coarse wire wheel or brush attached to a drill can also be used to remove paint that is over spot welds. Scrape off thick portions of undercoating or wax sealer before trying to remove paint.

Try to avoid using an oxyacetylene or propane torch to remove paint because it could overheat the metal. If you do use a torch, however, do not burn through the paint film so that the sheet metal panel begins to turn color. Heat the area only enough to soften the paint and then brush or scrape it off.

J. Plastic repair, like any other kind of body repair work, begins with the estimation process. It must be determined if the parts should be repaired or replaced. A minor crack, tear gouge, or hole in a nose fascia or large panel that is difficult to replace, costly, or not readily available probably indicates a repair should be considered. Extensive damage to the same component or damage to a fender extension or plastic trim item that is cheap and easy to replace would dictate replacement. In short, it is up to the repair person or estimator to decide if it makes more sense to repair a plastic part than to replace it.

If repair is the answer, it must be determined if the part needs to be removed from the vehicle. The entire damaged area must be accessible to do a quality repair. If it is not accessible, the part must be removed. Keep in mind that the part will also have to be refinished. Automotive plastics can generally be topcoated using conventional paint systems. Follow the manufacturer's recommendation to determine if a particular paint system can be used on a specific type of plastic, or if a special plastic primer or flex agent is required. Flexible plastics might require the addition of a flex agent to the paint system. The additive is needed because flexible plastics expand, contract, and bend easier than other substrates. Plastic parts are normally painted before they are installed.

K. Of primary concern to the technician should be vehicle safety items that need repair such as brake systems, steering and suspension systems, and restraint systems. The technician should examine the vehicle closely to verify any damage that is going to be repaired and that any item that was not damaged by the accident that may be worn and is unsafe is reported to the owner for repair. Another area of concern is environmental safety and the disposal of or reclaiming of refrigerants, lubricants, air bags, and catalytic converters; also the emission output of the vehicle should conform to the area's standards and requirements after repairs.

L. The first step in auto body repair is analyzing the damaged area. A number of conditions that the body technician must recognize are present in any damaged panel. The technician must look for direct damage, indirect damage, and work hardening whether normal or impact related.

Direct damage is the damaged portion of the panel that came in direct contact with the object that caused the impact. Indirect damage is caused by the shock of collision forces traveling through the body and of inertial forces acting on the rest of the unibody. Indirect damage can be more difficult to completely identify and analyze. It may be found anywhere on the vehicle. Work hardening is in all sheet metal panels of a vehicle to varying degrees. It is important to know where the metal was the hardest and softest before it was damaged.

Many times it may be best to draw the repair plan prior to actually pulling the vehicle. This drawing should show OEM actual dimensions, and anchoring and pulling locations. Before attempting any repair work, determine exactly the collision procedure that should be taken.

When it is determined how far the damage traveled in the unibody structure and when the damage is fully identified, the damaged area can be pulled and straightened. Repair the damage in the reverse (first-in, last-out) sequence to which it occurred during the collision. Plan the pulling sequence with the pulls in the opposite direction from those that caused the damage.

Signs of stress/deformation on unibody cars are:
- misaligned door, hood, trunk, and roof openings.
- dents and buckles in aprons and rails.
- misaligned suspension and motor mounts.
- damaged floor pans and rack and pinion mounts.
- cracked paint and undercoating.
- pulled or broken spot welds.
- split seams and seam sealer.

Stress-relieving uses hammer blows and sometimes carefully controlled heat to help return damaged metal to its original shape and state. A dolly or large wood block and hammer will work out a lot of stress. Most of the stress relieving will be "cold work." Not much heat will be used. If, however, heat is needed, control the heat carefully to prevent part damage and warpage. The best way to monitor heat application is with a heat crayon. Stroke or mark the cold piece with the crayon. When the stated temperature has been reached the crayon mark will liquefy.

M. It is often necessary to remove the paint film, undercoat, sealer, or other coatings covering body panel joints to find the location of the spot welds. To do this, remove the paint using a DA sander with medium grit paper or use a scuff-wheel in a grinder. A coarse wire wheel or brush attached to a drill can also be used to remove paint over spot welds. Scrape off thick portions of undercoating or wax sealer before trying to remove paint.

After the spot welds have been located, the welds can be drilled out and removed. Two types of cutting bits can be used: a drill type or a hole saw type. A plasma arc torch is seldom recommended for the removal of spot welds although it can be used. A high speed grinding wheel can also be used to separate spot welded panels. Use this technique only when the weld is not accessible with a drill, the replacement panel is on top, or a plug weld is too large to be drilled out.

After removal of the damaged panels, prepare the vehicle for installation of the new panels.

To do this, follow these steps:

1. Grind off the welding marks from the spot welding areas. Remove dirt, rust, paint, sealers, zinc coatings, and so on from joining surfaces and the back side of the joining surfaces.

2. Smooth the mating flanges with a hammer and dolly.

3. Apply weld-through primer to areas where the base metal is exposed after the paint film and rust have been removed from the joining surfaces.

To prepare the replacement panel for welding, follow these steps:

1. Use a disc sander to remove paint from both sides of the spot welding area.
2. Make plug holes for plug welding with a punch and a drill if necessary.
3. Apply weld-through primer to the welding surfaces where the paint film has been removed.
4. If the new panel is sectioned to overlap any of the existing panels, rough-cut the new panel to size.

Aligning new parts with the existing body is a very important step in body repair. Improperly aligned panels will affect both the appearance and the driveability of the repaired vehicle. Basically there are two methods of positioning body panels. With major damage, use dimension-measuring instruments to determine the right part position. With minor damage, you can often visually find the right panel position by the relationship between the new part and the surrounding panels.

N. The repair of aluminum panels requires much more care than the working of steel panels. Aluminum is much softer than steel, yet it is more difficult to shape once it becomes work hardened. It also melts at a lower temperature and distorts readily when heated.

It is important to keep in mind that aluminum body and frame parts are usually 1 1/2 to 2 times as thick as steel parts. When damaged, aluminum feels harder or stiffer to the touch because of work hardening. These characteristics must be taken into consideration when working with damaged aluminum panels.

O. The hood is the largest adjustable panel on most vehicles. It can be adjusted at the hinges, at the adjustable stops, and at the hood latch. The adjustments allow the hood to be moved up, down, forward, and rearward to align it with the fenders and cowl. The hood should align with the cowl and the fenders with an equal gap of approximately 5/32 inch (4 mm) between them. The front edge of the hood should be even with the front edge of the fender. When removing the hood, place a mark around the hinge to reference when installing the hood.

To adjust a hood, slightly loosen the bolts attaching the hood to the hinges. Keep them tight enough to allow you to shift the hood. Close the hood and line it up properly. Shift it by hand until the gaps around the sides of the hood are equal. Carefully raise the hood far enough for another technician to tighten the bolts. To right the adjustment of the hood up and down at the rear, slightly loosen the bolts holding the hinge to the fender or cowl. Then slowly close the hood and raise or lower its back side as necessary. When the back of the hood is level with the adjacent fenders and cowl, slowly raise the hood and tighten the bolts. After making the height and position adjustments, test the hood for proper latching. Slowly lower the hood and make sure it engages the latch in the center. If the hood must be slammed excessively hard to engage the latch, the latch should be raised. If the hood does not contact the front bump stops and they have already been adjusted, the latch should be lowered. If the hood bump stops are not contacting the hood, it could cause the hood to bounce or flutter when driving.

P. The trunk lid is very similar to the hood in construction. Two hinges connect the lid to the rear body panel. The trailing edge is secured by a latch. The trunk lid seals the trunk area from dust and water. Weather stripping is used to provide the proper seal. For the seal to be effective, the trunk lid must contact all of the weather stripping when the trunk lid is closed.

The lid must be evenly spaced between the adjacent panels. Slotted holes in the hinges and/or caged plates in the lid allow the trunk lid to be moved forward, rearward, and side to side. To adjust the lid forward or backward, slightly loosen the attaching hardware at both hinges. Close and adjust the trunk lid as required. Then raise the lid carefully and tighten the attaching hardware. In some cases, it might be necessary to use shims between the bolts and the trunk lid to raise or lower the front edges. If the front edge must be raised, place the shim(s) between the hinge and the lid in the front bolt area. To lower the front edge of the lid, place the shim(s) at the back of the hinge.

On some vehicles, the trunk lid has hinge assemblies that utilize torque rods to counterbalance the weight of the lid. This arrangement makes the trunk lid a lot easier to raise and hold it in the up position. The torque rods can be tightened or loosened by moving the torque rod end to a different hole or slot. Using a pipe inserted over the end of the torque rod is one way to safely move the rod to a new position.

The lock assembly is usually in the trunk lid, and the striker plate is bolted to the rear body panel. On some cars, this position is reversed. The trunk lid latch and striker can usually be adjusted up, down, and sideways to engage, align, and tightly hold the trunk lid.

Q. Door adjustment is needed so that the doors will close easily, not rattle, and not leak water and dust. Doors must fit their openings and align with the adjacent body panels. The door alignment is set by loosening the door hinge bolts and adjusting the door; shims can also be used. When making adjustments, it is a good idea to remove the door latch striker so that a false adjustment is not made. If door hinges are worn and sloppy, it is best to replace them before making any adjustments. When the door is adjusted properly, install the striker and adjust it. When the doors on a sedan need adjusting, start at the rear door. Since the quarter panel cannot be moved, the rear door must be adjusted to fit these body lines and the opening. Once the rear door is adjusted, the front door can then be adjusted to fit the rear door. Next, the front fender can be adjusted to fit the door. On hardtop models, the windows can then be adjusted to fit the weather stripping. The windows are usually adjusted starting with the front and working toward the back. The front is adjusted to fit the pillar, and the window is then adjusted to it. The rear door window is adjusted to the front window rear edge and the opening for the rear door assembly.

R. Vehicles with hatchback-type trunk lids are usually difficult to align because of their size. Many lids of this type are nearly horizontal in design, which makes them more prone to water and dust leaks. Some models use adjustable hinges. Others use welded hinges. The hatchback types also use gas filled door lift shock assemblies, or springs, one at each upper corner of the lid. Some play may not be available in the door lift support brackets to allow adjustment of the hatchback trunk lid.

Most late-model, full-size station wagons have a three-way tailgate. The three-way tailgate has a unique hinge and locking arrangement that allows the tailgate to be operated as a tailgate with the glass fully down or as a door. Before doing any station wagon tailgate alignment, closely examine the area to determine where the misalignment exists. It might be necessary to adjust the tailgate as a regular door. Closely examine the hinges to determine what adjustments are available. On some vehicles, the lower left hinge provides up/down as well as in/out adjustments for the tailgate. Some vehicles provide adjustment on the body side of the hinge. Others allow adjustment to the tailgate side of the hinge. Always refer to the vehicle's body repair manual for which hinges allow adjustment.

S. Bumpers are designed to protect the front and rear of the vehicle from damage during a low speed collision. Some bumpers are made of heavy gauge spring steel with a bright chromium metal. Other vehicles may have aluminum bumpers. Bumpers on may late-model cars are covered with urethane or other plastic. The use of urethane, polypropylene, or other plastics allows the bumper to be shaped to blend with the body contour. Plastic bumper covers can also be painted to match the body finish color. Underneath the plastic cover might be a steel or aluminum face bar or reinforcement bar, or a thick energy absorbing pad made of high density foam rubber or plastic. On older cars, bumpers were rigidly bolted to the vehicle's frame. At best, the old bumpers only resisted the bending forces of an impact; they transferred the energy of the shock directly to the frame. Manufacturers have since fitted their bumpers with energy absorbers. Most energy absorbers are mounted between the bumper face bar or bumper reinforcement and the frame.

There are many types of energy absorbers. The most common is similar to a shock absorber. The typical bumper shock is filled with hydraulic fluid. Upon impact, a piston filled with inert gas is forced into the cylinder. Under pressure, the hydraulic fluid flows into the piston through a small opening. The controlled flow of fluid absorbs the energy of the impact.

Another type of energy absorber is a thick urethane foam pad designed to rebound to its original shape in a mild collision. Some manufacturers use the bolts and brackets as energy absorbers. The bolts and brackets are designed to deform during a collision in order to absorb some of the impact force. The brackets must be replaced in most collision repairs.

Several cautions must be observed when removing bumpers with energy absorbers.

- The shock-type absorber is actually a small pressure vessel. It should never be subjected to heat or bending. If cutting or welding near an absorber, remove it.
- If the absorber is bound due to the impact, relieve the gas pressure before attempting to remove the bumper from the vehicle. Secure the bumper with a chain to prevent its sudden release and drill a hole into the front end of the piston tube to vent the pressure. Then remove the bumper and absorber.
- Work safely. Wear approved safety glasses when handling, drilling into, or removing a bound energy absorber.

Replacing a bumper is basically a matter of removing the right bolts. This job is made easier if the bumper is supported. On some vehicles, stone deflectors, parking lights, windshield washer hoses, and other items must be disconnected before the bumper can be removed from the car.

After bolting the bumper in place, it must be adjusted so that it is an equal distance from the fenders and grille. The clearance across the top must be even. Adjustments are made at the mounting bolts. The mounting brackets allow the bumper to move up or down, side to side, and in and out. If necessary, shims can be added between the bumper and the mounting brackets to adjust the bumper alignment.

T. Fenders are bolted to the radiator core support, the inner fender panel in the engine compartment, and the cowl behind the door and under the car. When these bolts are loosened, the fender can be moved for adjustment. The curvature of the fender must match the shape of the front door's edge. Therefore, any door adjustments that need to be made should be performed before fender adjustments. The fender to door alignment can be made by shimming the fender bolts. The gap between the fender and hood/door should be no more than 3/16 inch (4.8 mm). The front of the fender and the hood should be aligned as well. The result will be even spacing all around the fender and hood.

U. The doors are attached to the body with hinges. The hinges can be bolted to the door and body, or welded to either the body or door. No adjustment can be made to the welded door side hinge. A body-side hinge, however, can be adjusted forward, rearward, up, and down. The use of shims also allows the hinge to be moved in or out as desired.

To adjust a door, follow these steps:

1. Remove the striker bolt so it will not interfere with the alignment process.
2. Determine which hinge bolts must be loosened to move the door in the desired direction.
3. Loosen the hinge bolts just enough to permit movement of the door with a padded pry bar or jack and wooden block. On some vehicles, a special wrench must be used to loosen and tighten the bolts.
4. Move the door as needed. Tighten the hinge bolts. Then check the door fit to be sure there is no bind or interference with the adjacent panel.

5. Repeat the operation until the desired fit is obtained.

6. Install the striker bolt and adjust it so the door closes smoothly and flush with the rear door or quarter panel. Check that the door is in the fully latched position and not in the safety latch position.

7. On all hardtop models, the door and quarter glass must be checked to ensure proper alignment to the roof rail weather strip.

Worn door hinges will have play that allows up-and-down movement of the rear of the door. If the hinge pins are worn out, you should replace the hinge. Some hinges use bushings around the hinge pins. When these bushings are worn out, replace them. They will retighten the pin in the hinges and also readjust the door to a certain extent. Make sure replacement hinge bushings are available.

V. The actual work on the metal begins with the rough-out stage. "Rough-out" means to remove the most obvious damage to get the original part shape. It must be done properly if finishing operations are to succeed. Poor rough-out always costs the body technician money in time lost. Often in this situation, the technician hits up low areas and beats in all high ones thinking that eventually the metal will become straight. A body technician with a clear understanding of damage analysis knows metal is not straightened that way. Every buckle has a definite method of correction.

When the area has been bumped and pulled as level and smooth as possible, use a body file to locate any reaming high and low spots. File across the damaged area to the undamaged metal in the opposite side. The scratch pattern created by the file on the metal identifies any high and low spots. You can then further work the metal into shape before using plastic fillers.

W. A weld is formed when separate pieces of material are fused together through the application of heat. The heat must be high enough to cause the softening or melting of the pieces being joined. The three types of heat joining in collision repair are fusion welding, pressure welding, and adhesive bonding. Fusion welding is joining different pieces of metal together by melting and fusing them into each other. The pieces of metal are heated to their melting point, joined together (usually with filler rod), and allowed to cool. Pressure welding uses high heat and clamping pressure to join metals. No filler material is added to the weld. Current flow through the metals produces enough heat to melt the base metals. Adhesive bonding uses oxyacetylene to melt a filler metal onto the work piece for joining. Brazing is a form of adhesive bonding. It weakens steel by applying too much heat and is not recommended for collision repair.

Prepare the surface for welding by removing paint, undercoating, rust, dirt, oil, and grease. Use a plastic woven pad, sander, blaster, or wire brush. Avoid removing galvanized coatings. Apply weld-through primer to all bare metal mating surfaces. The surface to be welded must be bare, clean metal. If not, contaminants will mix with the weld puddle and may result in a weak, defective weld.

Locking pliers, C-clamps, sheet metal screws, and special clamps are all necessary tools for welding. Clamping both sides of a panel is not always possible. In these cases, a simple technique using self-tapping sheet metal screws or rivets can be employed. Fixtures can also be used in some cases to hold panels in proper alignment. Fixtures alone, however, should not be depended upon to maintain tight clamping forces at the welded joint. Some additional clamping will be required to make sure that the panels are tight together and not just held in proper alignment.

X. During a collision, the protective coatings on a vehicle are damaged. This occurs not just in the areas of direct impact but also in the indirectly damaged areas. Seams pull apart, caulking breaks loose, and paint chips and flakes. Locating the damage and restoring the protection to all affected areas remains a key challenge for the collision repair technician.

Anti-corrosion protection can be broken down into four categories:

1. Anti-corrosive compounds are either wax- or petroleum-based products resistant to chipping and abrasion. They can undercoat, deaden sound, and completely seal the surface. They should be applied to the underbody and inside body panels so

that they can penetrate into joints and body crevices to form a pliable, protective film.
2. Seam sealers prevent the penetration of water, mud, and fumes into panel joints. They serve the important role of preventing rust from forming between two adjacent surfaces.
3. Weld-through primers are used between the two pieces of base metal at a weld joint.
4. Rust converters change ferrous (red) iron oxide to ferric (black/blue) iron oxide. Rust converters also contain some type of latex emulsion that seals the surface after the conversion is complete. These products offer an interesting alternative for areas that cannot be completely cleaned.

Care is needed when applying anti-corrosion compounds. Keep the material away from parts that conduct heat, electrical parts, labels, identification numbers, and moving parts.

Y. Panel replacement can take one of two forms: replacement at the factory seams or sectioning. Replacement of panels along factory seams should be done when practical and economical. The result is almost identical to factory production in both strength and appearance. Sectioning involves cutting the part in a location other than the factory seam. This might or might not be a factory recommended practice. Sectioning a part should be analyzed to make sure it will not jeopardize structural integrity. Most manufacturers have specific recommendations for parts replacement. Always follow the procedures described in the body repair manual.

Z. Like other damaged panels, a door skin can be bumped back into shape, pulled into shape, or replaced. The decision is based on the amount of door frame damage. Another option is replacement of just the door skin. The door skin wraps around and is flanged around the door frame. The door skin is secured to the frame either by welding or with adhesives. Typical replacement procedures are needed for both types of skins. A welded skin has spot welds that hold the skin onto its frame. An adhesive door skin is bonded, not welded, to the door frame. It requires different replacement methods. To replace a door skin, follow the manufacturer's recommended procedures.

All the doors of unibody vehicles have inner metal reinforcements at various locations. There are also some other door frame reinforcements, such as at the hinge locations at the door lock plate. Door intrusion beams are normally used inside side doors. A door intrusion beam is welded or bolted to the metal support brackets on the door frame to increase the door strength.

AA. Plastic parts include bumpers, fender extensions, fascias, fender aprons, grille openings, stone shields, instrument panels, trim panels, fuel lines, and engine parts. Most of the new reinforced plastics are as strong as steel. This increasing use of plastic has resulted in new approaches to collision repair. Many plastic parts can be repaired more economically than being replaced, especially if the part does not have to be removed. Cuts, cracks, gouges, tears, and punctures are all repairable.

Reinforced plastic panels are normally bonded over a steel unibody structure. Reinforcements are also located behind the plastic panels. Plastic panel replacement or sectioning will depend on the amount and location of the damage. With proper backing strips used to reinforce the joints, sectioning can be done almost anywhere. There is sometimes a horizontal panel reinforcement. It can be made of reinforced plastic and is bonded to the back of the panel. This reinforcement contains mill and drill pad bolts, and acts as a spacer to hold the panel in its right place. When replacing a panel, use this reinforcement as a sectioning point and anchoring point for the new panel. After removing the scrap panel, prepare the space-frame for the new panel. First remove the old adhesives, and then bevel the outside edges of the existing panel to a 20 degree taper. Sand and clean the backsides of the panel where the backing strips will be attached. Remove the paint from those places where adhesive will be applied. Measure the replacement panel for fit. Trim the panel to size. Leave a gap between the existing panel and the replacement panel. Bevel the edge of the new panel, just like the existing panel. Apply the adhesive material in a continuous bead all of the way around the panel. Then install the panel onto the vehicle and clamp it into place.

AB. Some manufacturers place foam inside panels. Foam fillers are used to add rigidity and strength to structural parts. They also reduce noise and vibrations. Cutting and welding will damage the foam. Replacing the foam fillers must be a part of the repair procedure. The use and location of foam fillers varies from vehicle to vehicle. Follow the manufacturer's recommendations for replacing or sectioning foam filled panels. Some OEM replacement parts come with the foam already in the part. When the parts come without foam filler or when foam filler needs to be replaced, a product designed specifically for this application must be used to fill the panel.

Body panel joints and seams require special attention. As a general rule, a body sealant must be applied over all joints. The sealant must be applied so there are no gaps between the material and the panel surface.

There are four types of sealers that are commonly used in auto work:

1. Thin-bodied sealers are designed to fill seams under 1/8 inch (3.2 mm) wide. This sealer will shrink slightly to provide definition to the joint, while remaining flexible enough to resist vibration.

2. Heavy-bodied sealers are used to fill seams from 1/8 to 1/4 inch (3.2 to 6.4 mm) wide. These sealers can be tooled to hide the seam or can be left in bead form. Shrinkage should be minimal, with good resistance to sagging and high flexibility to resist cracking in service.

3. Brushable seam sealers are used on interior body seams where appearance is not important. These seams are normally hidden and not seen by the customer. These sealers are designed to hold brush marks and to resist salt and automotive fluids. Any seams such as those under the hood and under carriage that may be exposed to automotive fluids should have a brushable seam sealer.

4. Solid seam sealers containing 100 percent solids are used to fill larger voids at panel joints or holes. This product comes in strip caulking form and is designed to be pressed into place with your thumb.

When applying sealant, refer to the shop manual for the vehicle being repaired. Determine the sealer application area or look at the other side of the vehicle to see where the sealer is applied.

The bottom surfaces of the underbody and inside of the wheel house can be damaged by flying stones, which cause rust to develop. Therefore, the use of etching and conversion coating agents is of critical importance on exterior surfaces. Conversion coating provides the kind of superior paint film adhesion that retards creeping rust from working its way under the paint when chips and nicks do occur.

AC. Water leaks are noticed when moisture or rain enters the passenger compartment and collects on the carpeting. Air leaks normally cause a whistling or hissing noise in the passenger compartment during driving.

The principal methods used to locate air and water leaks are listed below:

- Spraying water on the vehicle
- Driving the vehicle over very dirty terrain
- Using a listening device
- Directing a strong beam of light on the vehicle and checking for light leakage between panels

Before making any actual leak test, remove all applicable interior trim from the general area of the reported leak. The spot where dust or water enters the vehicle might be some distance from the actual leak. Therefore, remove all trim, seats, or floor mats from the areas that are suspected as possible sources of the leak. Entrance dust is usually noticed at the point of entrance. These points should be sealed with an appropriate sealing compound and then rechecked to verify that the leak is sealed. Plugs and grommets are used in floor pans, dash panels, and trunk floors of a vehicle to keep dust and water from the interior. These items should be carefully checked to ensure they are in good condition.

Rattles and squeaks are sometimes caused by sheet metal that is too loose or rubbing adjacent parts. They are also caused by loose bolts and screws, and improperly adjusted doors, hoods, or body panels. Other rather simple things, such as a broken or loose exhaust mount, an improperly secured jack or tire, or articles in the trunk, can also cause a rattle.

Oftentimes, a noise will be pinpointed by the customer to be in a certain area of the vehicle when in fact it might be caused by something in another area of the vehicle. This is caused by sound traveling through the body. Usually thorough investigation and a test drive of the vehicle is recommended so that the rattle or noise can be located.

Most rattle noise repairs involve readjustment or replacement of parts, tightening loose attaching hardware, and welding broken parts. Many areas on the body of the vehicle can also cause rattles, noises, and squeaks. The most susceptible areas are the dash, doors, steering column, and seat tracks. All attaching hardware should be checked for tightness, especially in the area of the suspected noise source.

AD. Every vehicle has a variety of moldings. Moldings enhance the appearance of a vehicle by hiding panel joints, framing windshield or back lights, and accenting body lines. They also help to weatherproof by channeling wind and water away from window and doors. Moldings often must be replaced due to collision damage, or they can be added as a custom accessory. Moldings are fastened by adhesives or clips. The moldings can be installed after painting and when the vehicle is almost complete. The vehicle should be parked on a level surface and the moldings should be installed on the outermost surface of the vehicle. If installing factory molding refer to the vehicle service manual or the other side of the vehicle for proper placement.

Various pieces of trim are used in the passenger compartment for appearance and safety. Most are held on by a small clip or small screws. Sometimes screw heads can be covered by small plastic plugs. Screws can also be hidden under protruding parts. Your service manual will give locations for the fasteners holding interior trim parts. Vacuum the interior of the vehicle carefully. Clean the seats, door panels, seat belts, and carpets. Carefully remove any overspray that may have been left on windows or chrome.

AE. Grind the area to remove the old paint. Remove the paint for 3 or 4 inches (76 to 102 mm) around the area to be filled. If filler overlaps any of the existing finish, the paint film will absorb solvents from the new primer and paint, destroying the adhesion of the filler, and the filler will lift, cracking the paint and allowing moisture to seep in under the filler. Rust will then form on the metal. Use a 24 or 36 grit grinding disc to remove the paint. The coarse grit removes paint and surface rust quickly, and also etches the metal to provide better adhesion. You can also blast off the paint to prevent any removal of metal. If applying filler over a metal patch, avoid hammering down excess weld bead. Grind it level with the surface.

AF. After grinding away the finish from the repair area, blow away the sanding dust with compressed air and wipe the surface with a tack rag to remove any remaining dust particles. One of the most important steps in applying body fillers is surface preparation. Begin by washing the repair area to remove dirt and grime. Then clean the area with a wax and grease remover to eliminate wax, road tar, and grease. Also wash brazed or soldered joints with soda water to neutralize the acids in the flux. Do not grind these areas before neutralizing the acids. Grinding simply drives the acids deeper into the metal.

AG. When an area is stretched, the grains of the metal are moved farther away from each other. The metal is thinned and work hardened. Shrinking is needed to bring the molecules back to their original position and to restore the metal to its proper contour and thickness. Before shrinking, dolly the damage area back as close to its original shape as possible. Then you can accurately determine whether or not there is stretched metal in the damage area. It will usually pop in and out if stretched. If it is stretched, you must shrink the metal.

A small spot in the center of a warped area is heated to a dull red. When the temperature rises, the heated area of the steel panel swells and attempts to expand outward toward the edges of the heated circle (the circumference). Since the surrounding area is cool and hard, the panel cannot expand so a strong compression load is generated. If heating continues, the stretching of the metal is centered in the soft red-hot portion, pressing it out. This causes it to thicken, thus relieving the compression load. If the red hot area is cooled while in this state, the steel will contract and the surface area will shrink to less than its area before heating.

A variety of pieces of welding equipment can be used to heat metal for shrinking. Attachments are available for spot and MIG welding equipment to transform them into shrinking equipment. The most commonly used tool, however, is the oxyacetylene torch with a #1 or #2 tip.

AH. Body fillers come in cans and in plastic bags. When in a plastic bag, hand pressure or a dispenser can be used to force the filler onto your mixing board. This keeps the filler perfectly clean. A mixing board is the flat surface (metal, glass, or plastic) that is used for mixing the filler and its hardener.

Mix the can of filler to a uniform and smooth consistency free of lumps. Using a paint shaker will save time if the filler has been on the shelf for a while. If the body filler is not stirred up thoroughly to a smooth and uniform consistency before use, the filler in the upper portion of the can will be to thin and the lower portion of the can will be too thick and very coarse or grainy.

Loosen the cap of the hardener tube to prevent the hardener from being air bound. Hardener kneading is done by squeezing its contents back and forth inside the tube. This will ensure a smooth paste-like consistency. If the hardener is kneaded thoroughly and remains thin and watery, you have spoiled hardener. Hardener can spoil if frozen or if stored too long. It should not be used because it has broken down chemically.

Numerous problems can occur from improper catalyzing (mixing) of cream hardener (filler catalyst) and filler. Before catalyzing, make sure the materials to be used are compatible. They should be manufactured by the same company and be recommended for each other.

Add hardener according to the proportions indicated on the can, usually 10 percent hardener. Too little hardener will result in a soft, gummy filler that will not adhere properly to the metal. It will also not sand or feather-edge cleanly. Too much hardener will produce excessive gases resulting in pinholing.

With a clean putty knife or spreader, use a scraping motion to mix the filler and hardener together thoroughly and achieve a uniform color. Scrape filler off both sides of the spreader and mix it in. Every few strokes scrape the filler into the center of the mix board by circling inward. If the filler and hardener are not thoroughly mixed to a uniform color, soft spots will form in the cured filler. The result is an uneven cure, poor adhesion, lifting, and blistering.

AI. Apply the mixed filler promptly to a well-sanded and thoroughly cleaned surface. A tight, thin first application is recommended. Press firmly to force filler into sand scratches to maximize the bond. It is important to use the appropriate size spatula, or it will be difficult to apply a smooth layer of filler to the repair area. Rough filler takes extra time to sand off. Also make sure you move the spatula over the repair area to match the shape of the part.

When the filler is fully cured, you can apply additional coats as needed to build up the repaired area to the proper contour. Allow each application to set before applying the next coat of filler. Conventional body fillers should be built up slightly so that the waxy film that curing produces on the surface of the filler can be removed with a grater.

Applying the mixed filler thickly without first applying a thin application causes poor bonding and pinholing. Wiping over the repaired area with solvents before applying the mixed filler also causes pinholes and poor adhesion.

AJ. Allow the filler to cure to a semi-hard consistency. This usually takes fifteen to twenty minutes. If a firm white track is left when the filler is scratched with a fingernail, it is ready to be filed. Filing is perhaps the most important factor in achieving a quality surface and controlling material cost and labor. The file is used to cut excess filler to size quickly. Its long length produces an even, level surface. The teeth in the file are open enough to prevent the tool from clogging. Grinders, sanders, and air files do not level well; they become loaded quickly, create too much dust, and waste a lot of sandpaper.

After grading, sand out all the file marks with very coarse sandpaper. Use a block or air file on large flat surfaces. Use a disc orbital sander on smaller areas. Then follow with a finer 80 grit sandpaper until all grating scratches are removed. Final sanding should involve using 180 grit sandpaper until all 80 grit scratches are removed. The DA or air file can again be used or a long speed file can be used.

Be careful not to oversand. This results in the filled area being below the desired level, which makes it necessary to apply more filler. Oversanding is a common mistake for the novice body technician. Always sand a little and check your work. You want to slowly cut the filler down flush with the undamaged surface. On the flat surfaces, you can use a straightedge to check filler straightness. Run your hand over the area often to check for evenness. Do not be satisfied until the repaired surface feels perfectly even.

Another trick is to apply a guide coat or a thin mist of primer. By sanding off the primer guide coat, you can easily detect filler high and low spots. High spots will sand off more quickly. Low spots will leave the primer mist intact. When satisfied with the smoothness of the filler surface, clean the area with a tack cloth. A tack cloth picks up bits of filler dust that normal cleaning leaves behind. Remember that the tiniest particle will mar or ruin the paint job.

AK. Removing or servicing door glass methods will vary. Door window glass is secured in a channel with either bolts, rivets, or adhesive. Doors on sedans are basically the same, as are different makes of hardtop doors. Some hardtop doors require the removal of the upper window stops, lower lift brackets or bolts, the front or rear glass run channel, the upper stabilizers, and many other parts. If the glass is to be reinstalled, be sure to store it in a safe place.

Door glass requires servicing when it is broken or must be removed for other body or door repairs. The glass may also have to be removed to replace a broken channel assembly. On some vehicle makes, in order to remove the door glass, it may be necessary to remove the door trim panel, the water shield, the lower window stop, and the hardware securing the glass channel bracket to the glass channel. Mark the position of the channel on the glass. The sash channel can then be removed. To remove quarter glass, it might be necessary to remove some interior items, such as the rear seat, the window regulator handle, the trim panel(s), and the water shield.

If the window binds or is stiff, check the channel or add lubricant to the glass runs or guide channels. A door window that tips forward and that binds can be caused by improper adjustment of the lower sash brackets, a loose channel or cam roller, or a channel that is out of adjustment.

On some sedan doors, a full or partial length rubber glass run or channel is used. If the channel is too tight or lacks proper lubricant, the glass or rubber will bind. To free up the glass, a dry silicone spray should be applied to the glass run. Vehicle doors that use a full trim panel sometimes have a set of brackets at the top of the door. The trim panel is attached to these brackets. If they are set too far inward, the window glass will bind. Other items such as anti-rattle slides or other devices can cause the window glass to bind when raised. If not set correctly, they will cause binding.

If the door glass has to be adjusted to align with the edge of the quarter glass, be sure to check for proper quarter glass adjustment. Some of the quarter glasses are movable and some are stationary. To adjust or remove a quarter glass, some of the interior may have to be removed to gain access to the attaching mechanism. The manufacturer's specifications or procedures should be consulted for specific details.

The window regulator is a gear mechanism that allows you to raise and lower the glass. Regulators can be manual or powered electrically. The regulator and its associated parts are sometimes riveted to the door structure in lieu of being bolted. In this case, drill out the rivets in accordance with good shop practices, and reinstall the necessary parts using the appropriate rivet gun and rivets. For the regulators that are spot welded to the inner door panel, use a spot welder cutter to drill out the welds. If necessary, use a chisel between the regulator and inner panel to separate the two structures. Generally, the replacement is reinstalled with bolts.

AL. Many times the impact of an accident can cause indirect damage. Inspect the power sunroof panel for proper operation. Does it open and close easily without excessive noise from the motor? Check to see if it seals when it is closed. Check it for leaks. Most sunroofs have adjustments for the alignment of the roof panel glass. Usually bolts are used to fasten the glass and to adjust the panel. Consult the manufacturer's procedure for specific details.

AM. Inspect removable, manually operated roof panels, like T-tops, targa tops, or manual sunroofs for cracks, leaks, and broken hardware. Check to see if they are leaking and sealing when closed. Usually their latches or fasteners are adjustable. Consult the manufacturer's procedure for specific details.

AN. Water leaks are noticed when moisture or rain enters the passenger compartment and collects on the carpeting. Air leaks normally cause a whistling or hissing noise in the passenger compartment during driving. These leaks can sometimes be caused by damaged or ripped weather stripping. Weather stripping usually fits over a pinch weld flange or inside a channel. The rubber gasket can be glued on, held with screws or clips, or simply held securely by the design of the gasket. When applying weather stripping, cut the strip longer than required and butt the cut ends together. A sponge rubber plug is often used to hold the cut ends together. Some manufacturers require an application of silicone lubricant jelly at the base of the gasket. Be careful not to stretch the weather stripping during installation. Pulling the strip too tight will result in an improper seal.

AO. Inspect the convertible top for proper operation. Does it open and close easily without excessive noise from the motor? If replacing the top, check the fit. When the top is up, look for wrinkles and misalignment. Check to see if it seals when it is closed. Look at the linkage and the internal structure for damaged components. Check to see that the latches work and are adjusted correctly. Check it for wind and noise leaks. Most convertible tops have adjustments for the alignment in the front and back. Usually bolts are used to fasten the internal structure to the vehicle and to adjust the top. Consult the manufacturer's procedure for specific details.

AP. Stainless steel, steel, and aluminum are the only types of metals that can be welded in body repair. Check the repair area with a magnet to identify the material. If the magnet sticks it is steel. If the magnet does not stick, it could be aluminum or magnesium, (although they only produced magnesium parts for cars long ago). Aluminum looks similar to magnesium, which, if welded, could start a flash fire. To make sure the part is aluminum, brush the part with a stainless steel brush. Aluminum turns shiny; magnesium turns dull gray.

AQ. New welding techniques and equipment have entered the auto body repair picture, replacing the once-popular arc and oxyacetylene processes. New steel alloys used in today's cars cannot be welded properly by these two processes. Presently, gas metal arc welding (GMAW), better known as metal inert gas (MIG) welding, offers more advantages than other methods for welding components used in modern cars. Most of the applications of high strength, low alloy steel are confined to body structures, reinforcement gussets, brackets, and supports, rather than large panels or outer skin panels.

The advantages of MIG welding over conventional stick electrode arc welding are so numerous that manufacturers now recommend it almost exclusively. MIG welding is recommended by all OEMs, not only for high strength steel and unibody repair, but for all structural collision repair. This recommendation extends also to independent repair shops.

AR. Determining the right process depends on the manufacturer's recommendations. Refer to the manufacturer's body repair manual. Choose the right tool for the right procedure. (Are the parts to be welded aluminum? Should TIG welding be used? What thickness welding wire should be used? What type of shielding gas should be used?)

AS. Good welding results depend on proper arc length. The length of the arc is determined by the arc voltage. When the arc voltage is set properly, a continuous light hissing or cracking sound is emitted from the welding area. When the arc voltage is low, the arc length decreases, penetration is deep, and the bead is narrow and dome shaped. A sputtering sound and no arc means that the voltage is too low.

The tip-to-base distance is also an important factor in obtaining good welding results. The standard MIG welding tip-to-base distance is approximately 1/4 to 5/8 inches (6.3–15.9 mm). If the tip-to-base distance is too long, the length of wire protruding from the end of the gun increases and becomes preheated, which increases the melting speed of the wire. Also, the shield gas effect will be reduced if the tip-to-base distance is too long. If the tip-to-base distance is too short, it becomes difficult to see the progress of the weld because it will be hidden behind the tip of the gun.

There are two methods: the forward or forehand method and the reverse or backhand method. With the forward method, the penetration is shallow and the bead is flat. With the reverse method, the penetration is deep and a large amount of metal is deposited. The gun angle for both should be between 10 and 30 degrees.

Precise gas flow is essential to a good weld. If the volume of gas is too high, it will flow eddies and reduce the shield effect. If there is not enough gas, the shield effect will be reduced. Adjustment is made in accordance with the distance between the nozzle and the base metal, the welding current, the welding speed and the welding environment. The standard flow volume is approximately 1 3/8 to 1 1/2 cubic inches (0.022 to 0.024 liters) per minute or 15 to 25 cubic feet (420 to 700 cubic liters) per hour.

If you weld at a rapid pace, the penetration depth and bead width decreases, and the bead is dome shaped. If the speed is increased even faster, undercutting (in which the weld surface is lower than the base metal) can occur. Welding at too slow a speed can cause burn-through holes. Ordinarily, welding speed is determined by base metal thickness and/or the voltage of the welding machine.

An even, high-pitched buzzing sound indicates the right wire-to-heat ratio producing a temperature of approximately 9,000°F (4,982°C). Visual signs of the right setting occur when a steady reflected light starts to fade in intensity as the arc is shortened and wire speed is increased. If the wire speed is too low, a hiss and a plop sound will be heard as the wire melts away from the puddle and deposits the molten glob back. The visual result will be a much brighter reflected light. Too much wire speed will choke the arc; more wire is being deposited than the heat and puddle can absorb. The result is spitting and sputtering as the wire melts into tiny balls of molten metal that fly away from the weld. The visual signal is a strobe light arc effect.

AT. A welding helmet or welding goggles with proper shading lens must be worn when welding. These will protect the eyes and face from flying pieces of molten steel and harmful light rays. Sunglasses are not adequate protection. A welding filter lens, sometimes called a filter plate, is a shaded glass welding helmet insert for protecting your eyes from ultraviolet burns. The lenses are graded with numbers, from 4 to 12. The higher the number is, the darker the filter. The American Welding Society (AWS) recommends Grade 9 or 10 for MIG welding steel. There are also self-darkening lenses available that instantly turn dark when the arc is struck. There is no need to move the face shield up and down.

A respirator should also be used when welding galvanized metals. The fumes can cause serious lung or respiratory illness. Welding gloves should always be used when performing any welding operation to protect your hand from welding splatter and the heat of the welding operation, and to avoid cuts or abrasions. Sheet metal can cut your skin like a knife.

AU. Check the manufacturer's manual before hooking up equipment. Handle gas cylinders of shielding gas or welding gas with care. They might be pressurized to more than 2,000 pounds per square inch (13,790 kPa). Chain or strap the cylinder to a support sturdy enough to hold it securely to the MIG machine. Install the regulator, making sure to observe the recommended safety precautions. Never exceed recommended pressures when setting regulator pressures.

AV. When the welding machine's cable clamp is attached to clean metal on the vehicle near the weld site, it completes the welding circuit from the machine to the work and back to the machine. This clamp isn't really, as it is commonly referred to, a ground cable or ground clamp. The ground connection is for safety purposes and is usually made from the machine's case to the building ground through the third wire in the electric input cable. Locking jaw pliers, C-clamps, sheet metal screws, tack welds, or special clamps are necessary tools for good welding practices. Anybody can clamp panels together, but clamping panels together correctly to guarantee a sound weld will require close attention to every detail.

AW. Flat welding means the pieces are parallel with the bench or shop floor. Flat welding is generally easier and faster, and allows for the best penetration. When welding a member that is off the car, try to place it so that it can be welded in the flat position.

Horizontal welding has the pieces turned sideways. Gravity tends to pull the puddle down the joint. When welding a horizontal joint, angle the gun upward to hold the weld puddle in place against the pull of gravity.

Vertical welding has the pieces turned upright. Gravity tends to pull the puddle down the joint. When welding a vertical joint, the best procedure is usually to start the arc at the top of the joint and pull downward with a steady drag.

Overhead welding has the pieces turned upside down. Overhead welding is the most difficult. In this position, the danger of having too large a puddle is obvious; some of the molten metal can fall down into the nozzle, where it can create problems. Thus, always do overhead welding at a lower voltage while keeping the arc as short as possible and the weld puddle as small as possible. Press the nozzle against the work to ensure that the wire is not moved away from the puddle. It is best to pull the gun along with a steady drag.

AX. The last thing a technician wants to do when a vehicle comes into the shop for collision repair is create problems. This is especially true when it comes to electrical systems and electronic components. There are proper ways to protect automotive electrical systems and electronic components during storage and repair:
- Disconnect the battery cables before doing any kind of welding. The ground connection must be as close as possible to the work area to avoid a current seeking its own ground. Be careful about the placement of the welding cables. Do not let the welding cables run close to electronic displays or computers.
- Static electricity can cause problems. Avoid it by grounding yourself before handling any displays or electronic equipment.
- Avoid touching bare metal contacts. Oils from your skin can cause corrosion and poor connections.
- When replacing sensor wiring, always check the service manual and follow the routing instructions. Reuse or replace all electrical shielding; if not done, electronic crossover from the current carrying wires can affect the sensing and control circuits.
- Remove any computer that could be affected by welding, hammering, grinding, sanding, or metal straightening. Be sure to protect the removed computer and its connectors by wrapping them in plastic anti-static bags to shield them from moisture and dust.

AY. Prepare the surface for welding: remove paint, undercoating, rust, dirt, oil, and grease. Use a plastic woven pad, finder, sander, blaster, or wire brush. Avoid removing galvanized coatings. Apply weld-through primer to all bare metal mating surfaces. The surface to be welded must be bare, clean metal. If not, contaminants will mix with the weld puddle and may result in a weak, defective weld. If the new panel is sectioned to overlap any of the existing panels, rough-cut the new panel to size.

Aligning new parts with the existing body is a very important step in body repair. Improperly aligned panels will affect both the appearance and the driveability of the repaired vehicle. Basically there are two methods of positioning body panels. With major damage, use dimension measuring instruments to determine the right part position. With minor damage, you can often visually find the right panel position by the relationship between the new part and the surrounding panels.

Locking pliers, C-clamps, sheet metal screws, and special clamps are all necessary tools for welding. Clamping both sides of a panel is not always possible. In these cases, a simple technique using self tapping sheet metal screws or rivets can be employed. Fixtures can also be used in some cases to hold panels in proper alignment. Fixtures alone, however, should not be depended upon to maintain tight clamping forces at the welded joint. Some additional clamping will be required to make sure that the panels are tight together and not just held in proper alignment.

AZ. Butt welds are formed by fitting two edges of adjacent panels together and welding along the mating or butting edge of the panel. In butt welding especially thin panels, it is wise not to weld more than 3/4 inch (19 mm) at one time to prevent panel warpage from heat.

Lap and flange welds are made with identical techniques. They are formed by welding or fusing two surfaces to be joined at the edge of the top one of two overlapping surfaces. These are similar to butt welds except only the top surface has an edge. Lap and flange welds should be made only in repairs where they replace original factory lap or flange welds, or where outer panels and nonstructural panels are involved. These welds should not be used to join more than two thicknesses of material together.

The plug weld is the body shop alternative to the OEM resistance spot weld made at the factory because it can be used anywhere in the body structure that the factory used a resistance spot weld. Its use is not restricted. It has ample strength for welding load-bearing structural members. It can also be used on cosmetic body skins and other thin gauge sheet metal. Plug welding is a form of spot welding. A plug weld is formed by drilling or punching a hole in the outer panel being joined. The material should be tightly clamped together. Holding the gun at right angles to the surface, aim the electrode wire in the hole, and trigger the arc while moving the gun in a circular motion around the hole. The puddle fills the hole and solidifies.

BA. For MIG spot welding, a special welding nozzle must replace the standard nozzle. Once in place, and with the spot timing, welding heat, and backburn time set for the given situation, the spot nozzle is held against the weld site and the trigger. For a very brief period of time, the timed pulses of wire feed and welding current are activated, during which the arc melts through the outer layer and penetrates the inner layer. After this, the automatic shutoff goes into action and no matter how long the trigger is squeezed, nothing will happen. The trigger must be released and then squeezed again to obtain the next spot pulse. Because of varying conditions, the quality of a MIG spot weld is difficult to determine. Therefore, MIG plug welding is the preferred method of welding load-bearing members.

The MIG lap spot technique is a popular one for the quick, effective welding of lap joints and flanges on thin gauge nonstructural sheets and skins. Here again the spot timer is set, but this time the spot nozzle is positioned over the edge of the outer sheet at an angle slightly off 90 degrees. This will allow contact with both pieces of metal at the same time. The arc melts into the edge and penetrates the lower sheet.

In MIG stitch welding, the standard nozzle is used, not the spot nozzle. To make a stitch weld, combine spot welding with a continuous welding technique. To do this, set the automatic shutoff timer or pulsed interval timer, depending on the machine. The spot weld pulses and shutoff occurs with automatic regularity; weld then stop, weld then stop, weld then stop as long as the trigger is held in. The arc off period allows the last spot to cool slightly and start to solidify before the next spot is deposited. This intermittent technique means less distortion and less melt-through or burn-through. These characteristics make the stitch weld preferable to the continuous weld for working thinner gauge cosmetic panels.

BB. Proper welding techniques ensure good welding results. If welding defects should occur, think of ways to change your procedure to right the defect. When making any MIG repairs, the materials and panels must be similar enough to allow mixing when they are welded together. The combination of the cleanliness of the welded area, the mixing of proper metals, and the right heat application will result in good MIG welds. A welding problem causes a weak or cosmetically poor joint that reduces quality.

Some common weld problems include:
- weld porosity - (holes in the weld).
- weld cracks - (cracks on the top or inside the weld bead).
- weld distortion - (uneven weld bead).
- weld spatter - (drops of electrode on and around the weld bead).
- weld undercut - (groove melted along either side of the weld and left unfilled).
- weld overlap - (excess weld metal mounted on top and either side of the weld bead).
- too little penetration - (weld bead sitting on top of the base metal).
- too much penetration - (burn-through beneath the lower base metal).

BC. Plasma arc cutting creates an intensely hot air stream over a very small area that melts and removes metal. Extremely clean cuts are possible with plasma arc cutting. Because of the tight focus of the heat, there is no warpage, even when cutting thin sheet metal. Plasma arc cutting is replacing oxyacetylene as the best way to cut metals. It cuts damaged metal effectively and quickly but will not destroy the proper ties of the base metal. The old method of flame cutting just does not work that well anymore.

In plasma cutting, compressed air is often used for both shielding and cutting. As a shielding gas, air covers the outside area of the torch nozzle, cooling the area so the torch does not overheat. Air also becomes the cutting gas. It swirls around the electrode as it heads towards the nozzle opening. The swirling action helps to constrict and narrow the gas. When the machine is turned on, a pilot arc is formed between the nozzle and the inner electrode. When the cutting gas reaches this pilot arc, it is superheated up to 60,000°F (33,315°C). The extreme heat force of the cutting arc melts a narrow path through the metal. This serves to dissipate the metal into gas and tiny particles. The force of plasma literally blows away the metal particles, leaving a clean cut.

A 10–15 amp plasma arc cutter is generally adequate for mild steel up to 3/16 inch (4.8 mm) thick; a 30 amp unit can cut metal up to 1/4 inch (6.4 mm) thick, and a 60 amp unit will slice through metal up to 1/2 inch (12.7 mm) thick.

BD. Two general types of plastics are used in automotive construction: thermoplastics and thermosetting plastics.

Thermoplastics can be repeatedly softened and reshaped by heating, with no change in their chemical makeup. They soften when heated and harden when cooled. Thermoplastics are weldable with a plastic welder.

Thermosetting plastics undergo a chemical change by the action of heating, a catalyst, or ultraviolet light. They are hardened into a permanent shape that can not be altered by reapplying heat or catalysts. Thermosets are not weldable, but can be repaired with flexible part repair materials.

Composite plastics, or hybrids, are blends of different plastics and other ingredients designed to achieve specific performance characteristics.

A good example of the change in unibody vehicles centers around the use of fiber reinforced composite plastic panels, commonly known as sheet molded compounds (SMC). The reason for using an SMC is simple. It is light, corrosion-proof, dent-resistant, and relatively easy to repair compared to the more traditional materials. The use of SMC and other fiber reinforced plastics (FRP) is not new. They have been used in various applications on automobiles for years. The use of large external body panels of reinforced plastic is not unusual either. What is new is that, unlike the external panels on earlier vehicles, these panels now are bonded to a metal space-frame using structural adhesives, thus adding overall structural rigidity to the vehicle.

There are several ways to identify an unknown plastic. One way is by international symbols, or ISO codes, which are molded into plastic parts. Many manufacturers are using these symbols. The symbol or abbreviation is formed in an oval on the back side of the part. One problem is that you usually have to remove the part to read the symbol. If the body part is not identified by a symbol, the body repair manual will give information about plastic types used on the vehicle. Body manuals often name the types of plastic used in a particular application.

The burn test involves using a flame and the resulting smoke to determine the type of plastic. It is no longer recommended for identifying plastic. First, an open flame in a collision repair shop creates a potential fire hazard. Secondly, it is environmentally unsound. Finally, the burn test is not always reliable. Many parts are now being manufactured from composite plastics that use more than one ingredient. A burn test is of no help in such cases.

A reliable means of identifying an unknown plastic is to make a trial and error weld on a hidden or damaged area of the part. Try several different filler rods until one sticks. Most suppliers offer only a few types of plastic filler rods; the range of possibilities is not that great and the rods are color coded. Once you find a rod that works, the base material is identified.

BE. Plastic welding uses heat and sometimes a plastic filler rod to join or repair plastic parts. The welding of plastic is not unlike the welding of metals. Both methods use a heat source, welding rod, and similar techniques. Joints are prepared in much the same manner and evaluated for strength.

Hot air plastic welding uses a tool with an electric element to produce hot air that blows through a nozzle and melts plastic. The air supply comes from either the air compressor or a self-contained portable compressor that comes with the welding unit. The torch is used in conjunction with the welding rod, which is normally 3/16 inch (5 mm) in diameter. The plastic welding rod must be made of the same material as the plastic being repaired. One of the problems with hot air plastic welding is that the plastic welding rod is often thicker than the panel to be welded. This can cause the panel to overheat before the rod has melted. Using a smaller diameter rod with the hot air welder can often right such warpage problems. Some hot air welder manufacturers have developed specialized welding tips and rods to meet specific needs. Check the product catalog for more information.

Airless plastic welding uses an electric heating element to melt a smaller diameter rod with no external air supply. It has become very popular. Airless welding with a smaller rod helps eliminate two troublesome problems: panel warpage and excess rod build-up. Make sure the rod is the same material as the damaged plastic, or the weld will be unsuccessful. Many airless welder manufacturers provide rod application charts. When the right rod has been chosen, it is good practice to run a small piece through the welder to clean out the tip before beginning.

Ultrasonic plastic welding relies on high frequency vibratory energy to produce plastic bonding without melting the base material. Hand held systems are available in 20 and 40 kHz frequencies. They are equally adept at welding large parts and tight, hard to reach areas. Welding time is controlled by the power supply. Most commonly used injection molded plastics can be ultrasonically welded without the use of solvents, heat, or adhesives. Ultrasonic weldability depends on the plastic's melting temperature, elasticity, impact resistance, coefficient of friction, and thermal conductivity. Generally, the more rigid the plastic, the easier it is to weld ultrasonically. Thermoplastics are ideal for ultrasonic welding, provided the welder can be positioned close to the joint area.

In the typical ultrasonic system, the vibration is generated in the transducer and then transmitted through the sonotrode, which is the equivalent of an electrode. The sonotrode tip directly contacts the work piece. An anvil supports the assembly. The best results are achieved when the sonotrode tip and anvil are contoured to accommodate the specific shape of the parts being joined.

BF. When welding plastic, single or double V-butt welds produce the strongest joints. When using a round or V-shaped welding rod, prepare the area by slowly grinding, sanding, or shaving the adjoining surfaces to produce a single or double V. Wipe any dust shavings from the joint with a clean, dry rag. Do not use cleaning solvents because they can soften the plastic edges and cause poor welds.

To weld a rigid plastic or flexible part, set the temperature on the welder for the plastic being welded. Allow it to warm up to the proper temperature. Clean the part by washing with soap and water, followed by a good plastic cleaner. Align the break using aluminum body tape. V-groove the damaged area 75 percent of the way through the base material. Angle or bevel back the torn edges of the damage at least 1/4 inch (6.4 mm) on each side of the damaged area. Use a die grinder or similar tool. Clean the preheated tube, and insert the rod. Begin the weld by placing the shoe over the V-groove and feeding the rod through. Move the tip slowly for good melt-in and heat penetration. When the entire groove has been filled, turn the shoe over and use the tip to stitch tamp the rod and base material together into a good mix along the length of the weld. Smooth the weld area using a flat shoe part of the tip, again working slowly. Then cool with a damp sponge or cloth. Shape the excess weld build-up to a smooth contour, using a razor blade and/or abrasive paper.

BG. Adhesive repair systems are of two types: cyanoacrylates (CA) and two-part. Two-part is the most commonly used. CAs are one-part fast curing adhesives used to help repair rigid and flexible plastics. They are used as a filler or to tack parts together before applying the final repair material. CAs are sometimes known as super glues. Two-part adhesive systems consist of a base resin and a hardener. The resin comes in one container and the hardener in another. When mixed, the adhesive cures into a plastic material similar to the material in the part. Two-part adhesive systems are an acceptable alternative to welding for plastic repairs. Not all plastics can be welded, but adhesives can be used in all but a few instances.

To repair a rigid or flexible part with two-part adhesive, first clean the part with soap and water and then a good plastic cleaner. Make sure both the part and the repair material are at room temperature for proper curing and adhesion. Mix the two parts of the adhesive thoroughly and in the proper proportions. Mix until the color is uniform. Apply the material within the time guidelines given in the product literature. Use heat if indicated by the manufacturer. Follow the cure time guidelines given in the product literature. Regulated heat can speed curing. Support the part adequately during the cure time to ensure that the damage area does not move before that adhesive cures. This would weaken the repair. Follow the product literature for guidelines on when to reinforce a repair.

BH. An adhesion promoter is a chemical that treats the surface of the plastic so the repair material will bond properly. Some plastics require an adhesion promoter. Lightly sand a hidden spot on the piece with a high speed grinder and a 36 grit paper. If the material gives off dust, it can be repaired with a standard structural adhesive system. If the material melts and smears or has a greasy or waxy look, then you must use an adhesion promoter.

Here is a typical way to use a two-part adhesive to repair a flexible part. Clean the surface with soap and water. Wipe or blow dry. Then clean the surface with a good plastic cleaner. V-groove the damaged area. Then grind about a 1 1/2 inch (38 mm) taper around the damaged area. Then blow off the dust. To reinforce the repair area, sand and clean the backside of the part with a plastic cleaner. Then, if needed, apply a coat of adhesion promoter. Dispense equal amounts of both parts of the adhesive. Mix them to a uniform color. Apply the material to a piece of fiberglass cloth using a plastic squeegee. Attach the plastic saturated cloth to the backside of the part. Fill in the weave with additional adhesive material. With the backside reinforcement in place, apply a coat of adhesion promoter to the sanded repair area on the front side. Let the adhesion promoter dry completely. Fill in the area with adhesive material. Shape the adhesive with your spreader to match the shape of the part. Allow it to cure properly. Rough grind the area, then sand it smooth. If additional adhesive material is needed to fill in a low spot or pinholes, be sure to apply a coat of adhesion promoter again.

BI. When welding textured plastics, care must be taken not to overheat the material or you could damage the textured surface. When the repair is finished, you still have to retexture the surface to match the grain of the existing surface. First make sure the contour of the repair is right. Then spray on a vinyl retexture material to match the existing surface. Blend it out to a break or contour line on the cover. This will help hide the repair. Reapply the material until an acceptable texture is made. The texture doesn't have to match the same pattern; as long as it has the same coarseness, it will blend in.

BJ. Vinyl is a soft, flexible, thin plastic material often applied over a foam filler. Vinyl over foam construction is commonly used on interior parts for safety. Common vinyl parts are the dash pads, armrests, inner door trim, seat covers, and exterior roof covering. Dash pads or padded instrument panels are expensive and time-consuming to replace. Therefore, they are perfect candidates for repair. Most dash pads are made of vinyl-clad urethane foam to protect people during a collision. Surface dents in foam dash pads, armrests, and other padded interior parts are common in collision repair. These dents can often be repaired as follows:

1. Soak the dent with a damp sponge or cloth for about half a minute. Leave the dented area moist.
2. Using a heat gun, heat the area around the dent. Hold the gun 10 to 12 inches (254 to 305 mm) from the surface. Keep it moving in a circular motion at all times, working from the outside in.
3. Heat the area around to 130°F (54°C). Do not overheat the vinyl or it will blister. Keep heating until the area is too uncomfortable to touch. If available, use a digital thermometer to meter the surface temperature.
4. Using gloves, massage the pad. Force the material toward the center of the dent. The area might have to be reheated and massaged more than once. In some cases, heat alone might repair the damage.
5. When the dent has been removed, cool the area quickly with a damp sponge or cloth.
6. Apply vinyl treatment or preservative to the part.

BK. Many bent, stretched, or deformed plastic parts such as flexible bumper covers and vinyl clad foam interior parts, can often be straightened with heat. This is because of plastic memory, which means the piece wants to keep or return to its original molded shape. If it is bent or deformed slightly, it will return to its original shape if heat is applied. To reshape a distorted bumper cover, use the following procedure:

1. Thoroughly wash the cover with soap and water.
2. Clean with plastic cleaner. Make sure to remove all road tar, oil, grease, and undercoating.
3. Dampen the repair area with a water soaked rag or sponge.
4. Apply heat directly to the distorted area. Use a concentrated heat source, such as a heat lamp or high-temp heat gun. When the opposite side of the cover becomes uncomfortable to the touch, it has been heated enough.
5. Use a paint paddle, squeegee, or wood block to help reshape the piece if necessary.
6. Quickly cool the area by applying cold water with a sponge or rag.

BL. Proper sectioning requires that you understand which areas are most appropriate for sectioning. You must also know how to avoid problems with horizontal bracing, rivets, and concealed parts. The replacement panel used will depend on the amount and location of the damage. Using the left rear quarter panel as an example, there are three possibilities. The entire panel can be ordered, or simply a front or rear half can be ordered. Remember that reinforced plastic is a very forgiving and workable material.

Just because quarter panels come split at the wheel well, the sectioning point does not have to be located there. With proper backing strips to reinforce the joints, sectioning can be done almost anywhere.

The mill and drill pads are used to help hold the factory panels in place while the adhesive cures. These mill and drill pads will also help you to hold, align, and level replacement panels. If a panel is to be sectioned, it should be done between mill and drill pad locations. First, remove the interior trim to locate the horizontal bracing and mill and drill pads. Examine the back of the panel to gauge the extent of panel damage. Also determine the location of the horizontal bracing, mill and drill pads, and electrical and mechanical components.

Once the interior trim is removed, an opening can be cut. Controlling the depth of the cut is very important. Space-frame components, as well as electrical lines and heating and cooling elements, may be located behind the panels. When cutting the opening, know what is behind the panel being cut into, or limit the cut to a depth of 1/4 inch (6 mm) to avoid doing damage. Now that the opening has been cut, the rest of the panel can be removed from the space-frame. This can be done by using heat and a putty knife, or by carefully using an air chisel. Choose a flat chisel, beveled on one side only. Be careful not to damage the space-frame. If the door surround panels are to be left attached to the vehicle, the air chisel method may not be the best choice for separating the seam between two panels. Use the heat and putty knife method to separate the seams to avoid doing damage to the door surface pieces.

Sometimes there is a horizontal panel reinforcement. It can be made of reinforced plastic and is bonded to the back of the panel. This reinforcement contains the mill and drill pad bolts, and acts as a spacer to hold the panel in its right place. When replacing a panel, use this reinforcement as a sectioning point and anchoring point for the new panel. Leave several inches of the reinforcement bonded to the space-frame. Make the cut, using extra care to avoid cutting through the reinforcement. Decide how much of the reinforcement will be kept. Make a cut on the scrap side of the panel. Cut through both the panel and reinforcement. Remove the scrap panel in the usual way. Then remove the strip of panel left attached to the reinforcement.

On the replacement panel, mark off the corresponding piece of panel reinforcement that must be removed to fit the panel. Be cautious when estimating how much reinforcement must be removed from the replacement panel. Leave enough reinforcement to enable the part to be trimmed to fit.

BM. A molded core is a curved body repair part made by applying plastic repair material over a part and then removing the cured material. Naturally, holes are much more difficult to repair in a curved portion of a reinforced plastic panel than those on a flat surface. Basically, the only solution (short of purchasing a new panel section) is to use the molded core method of replacement. This is often the quickest and cheapest way to repair a curved surface.

Locate an undamaged panel on another vehicle that matches the damaged one. It will be used as a model or pattern. A new or used vehicle can be used since it will not be harmed.

In some instances, it might not be possible to place the core on the inside of the damaged panel. In this case, the damaged portion must be cut out to the exact size of the core. After the panel has been trimmed and its edges beveled, tabs must be installed to support the core from the inside. These tabs can be made from pieces of the panel or from fiberglass strips saturated in resin/hardener.

After cleaning and sanding the inside sections, attach the tabs to the inside edge of the panel and bond with fiberglass adhesive. Clamping pliers can be used to hold the tabs in place. Taper the edge of the opening and place the core on the tabs. Fasten the core to the tabs with fiberglass adhesive. Grind down any high spots so that layers of fiberglass mat can be added.

Place the saturated mats over the core. Work each layer with a spatula or squeegee to remove all air pockets. Additional resin/hardener can be added with a paintbrush to secure the layers. Allow sufficient curing time. Then sand the surface level for a smooth surface; use fiberglass filler to finish the job.

BN. After removing the scrap panel, prepare the space-frame for the new panel. First remove the old adhesive from the space-frame. You may have to use heat and a putty knife or a sander. Bevel the outside edges of the existing panel to a taper. Sand and clean the backsides of the panel where the backing strips will be attached. Backing strips are made using scrap material that duplicates the original panel contour as closely as possible. They should extend beyond either side of the sectioning location. Clean the backing strips. Remove the paint from those places where adhesive will be applied.

Measure the replacement panel for fit. Trim the panel to size; leave a gap between the existing panel and the replacement panel. When a proper fit has been established, the new panel can be prepared for the adhesive. Sand or grind bevels into the panel where they mate to the existing panel. Bevel the mating edges of the new panel to a shallow taper, just like on the existing panel. Make sure to bevel all of the way through the panel. Do not leave a shoulder. Apply adhesive material in a continuous bead all the way around the panel. Check for horizontal bracing, and apply a bead of adhesive to correspond with it. Then fit the panel onto the vehicle and clamp it into place. Install the mill and drill pad nuts, and tighten them securely. The work life will be recommended by the adhesive manufacturer. This time allowance should be followed to ensure that the proper fit has been achieved.

BO. Reinforced reaction injection molded (RRIM) is a two-part polyurethane composite plastic. Part A is the isocyanate. Part B contains the reinforcing fibers, resins, and a catalyst. Two-parts are first mixed in a special mixing chamber, then injected into a mold. RRIM parts are becoming more common in fenders and bumper covers. Since RRIM is a thermosetting plastic, heat is applied to the mold to cure the material. The molded product is made to be stiff yet flexible. It can absorb minor impacts without damage. This makes RRIM an ideal material for exposed areas. Gouges and punctures can be repaired using a structural adhesive. If the damage is a puncture that extends through the panel, a backing patch is required.

BP. Over one hundred types of plastic are currently being used in the manufacture of vehicles, and approximately forty need preparation before painting. The painter must be able to identify these plastic parts before refinishing them. If the parts are factory primed, no additional priming is necessary; if they are not, parts might benefit from the use of a special plastic primer or primer sealer to improve paint adhesion.

Exterior, hard rigid parts should be treated as fiberglass when in doubt as to their makeup. In fact, fiberglass should be treated much the same in preparation of the final coat as body steel. It must be remembered that fiberglass parts do not require chemical conditioners. Replacement or new panels can contain contaminants on the surface due to the release agents used in the molds. Several common release agents are composed of silicone oils. These contaminants must be removed. Newly molded parts should be washed with denatured alcohol used liberally on a clean cloth. Thoroughly clean the surface with an approved material. Sand exposed fiberglass by hand or with a sander. Reclean the surface and wipe dry with clean rags. When refinishing previously painted fiberglass parts, care should be taken not to sand through the gel coat, and sealer should be used. Fiberglass parts are extremely porous. The gel coat keeps topcoat solvents from being absorbed into the substrate.

When finishing a previously painted sheet molded compound (SMC) with either a blend or full panel paint procedure, it is necessary to apply a coat of an adhesion promoter. This must be applied beyond the blend area when performing a spot repair. In the event of refinishing a full panel, the entire part must be coated. A flash time of at least thirty minutes is required before applying the base color. This will ensure adequate adhesion of the topcoat.

Sample Test for Practice

Sample Test

Please note the letters in parentheses following each sample question. These letters match the overview in section 4 that discusses the relevant subject matter. You may want to refer back to the overview using this cross-referencing key to help with questions posing problems for you.

1. Technician A says rough filler takes extra time to sand off.
   Technician B says filler should be applied thickly on the first application.
   Who is right?
   A. A only
   B. B only
   C. Both A and B
   D. Neither A nor B (AI)

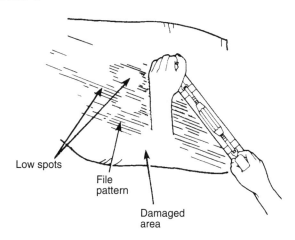

2. In the figure above, after an area has been bumped and pulled as level as possible, you locate any remaining high or low spots by using a:
   A. guide coat.
   B. body file.
   C. straight edge.
   D. grinder. (V)

3. Technician A says that a special shielding gas must be used to spot weld.
   Technician B says some MIG welders can be used for spot welding.
   Who is right?
   A. A only
   B. B only
   C. Both A and B
   D. Neither A nor B (BA)

59

4. Moldings are used on vehicles for all of the following reasons EXCEPT:
   A. they enhance appearance.
   B. they channel wind and water away from the windows and doors.
   C. they hide panel joints.
   D. they improve aerodynamics. (AD)

5. Technician A says that repairs should be started as soon as all the parts on the estimate have arrived.
   Technician B says even though the customer and insurance company recommend replacement, he is going to repair the quarter panel on the vehicle because it is possible.
   Who is right?
   A. A only
   B. B only
   C. Both A and B
   D. Neither A nor B (A)

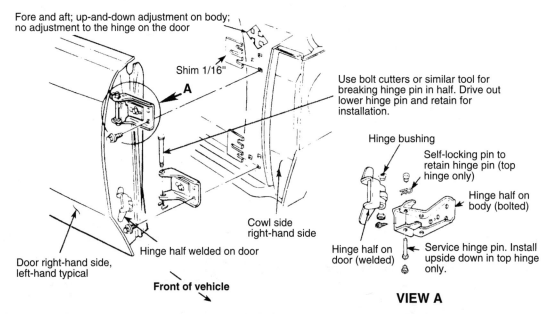

6. Technician A says welded hinges, as shown above, that are damaged are not replaceable.
   Technician B says that there is no adjustment on welded door hinges.
   Who is right?
   A. A only
   B. B only
   C. Both A and B
   D. Neither A nor B (U)

7. When roughing out an area, you are trying to:
   A. sand the area.
   B. prepare the area for fillers.
   C. restore a part to its original shape.
   D. remove the damaged panels. (V)

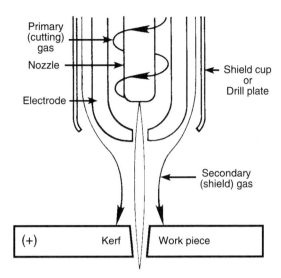

8. Technician A says plasma cutting, as shown above, causes too much distortion when cutting thin metals.
   Technician B says plasma arc cutting is being replaced by oxyacetylene welding.
   Who is right?
   A. A only
   B. B only
   C. Both A and B
   D. Neither A nor B   (BC)

9. Which of the following is not true?
   A. Cracked paint and undercoating is a sign of stress.
   B. Most of the stress-relieving will be "cold work."
   C. The best way to monitor heat application is with a heat crayon.
   D. Work hardened panels should be replaced.   (L)

10. Technician A says contaminants in the weld puddle will not affect the strength of the weld.
    Technician B says the surface to be welded must be bare.
    Who is right?
    A. A only
    B. B only
    C. Both A and B
    D. Neither A nor B   (W)

11. Too little weld penetration will cause the weld to:
    A. be even with the base metal.
    B. sit on top of the base metal.
    C. burn through the base metal.
    D. be uneven and warp the base metal.   (BB)

12. What is the best method for separating spot welds?
    A. Blowing them out with a plasma torch
    B. Cutting them out with a spot cutter
    C. Drilling them out
    D. Grinding them down with a high speed grinding wheel   (M)

13. What type of plastics can be repeatedly softened and reshaped by heating?
    A. Thermoplastics
    B. Thermosetting
    C. Composite
    D. Thermoactive (BD)

14. How many general types of plastic are mainly used in the automotive industry?
    A. 1
    B. 2
    C. 3
    D. 4 (BD)

15. Technician A says after lifting a vehicle to the desired height, always lower the hoist onto the mechanical safeties.
    Technician B says that there are many safe positions from which to lift a vehicle.
    Who is right?
    A. A only
    B. B only
    C. Both A and B
    D. Neither A nor B (B)

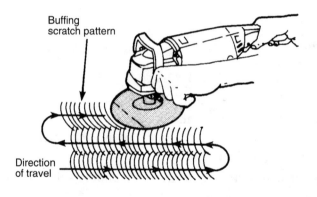

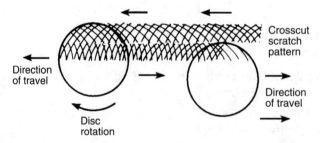

16. What grit grinding disc, as shown above, should be used to remove paint and surface rust quickly?
    A. 14 grit disc
    B. 36 grit disc
    C. 60 grit disc
    D. 80 grit disc (AE)

17. When working a metal patch with an excessive weld bead, you should:
    A. hammer it flat.
    B. heat-shrink it.
    C. grind it flat.
    D. reweld it. (AE)

18. Technician A says the proper gun angle when MIG welding is 90 degrees. Technician B says there is only one method of gun travel when MIG welding, the backhand method.
    Who is right?
    A. A only
    B. B only
    C. Both A and B
    D. Neither A nor B (AS)

19. A customer brings back a vehicle with a rattle coming from the back of the car.
    Technician A says the spare tire or jack might be loose.
    Technician B says the sound could be traveling through the body and the rattle might be up front.
    Who is right?
    A. A only
    B. B only
    C. Both A and B
    D. Neither A nor B (AC)

20. What is the right sound emitted from the weld area when MIG welding correctly?
    A. A continuous light hissing or cracking
    B. An intermittent sputtering sound
    C. A hiss and a plop sound
    D. A continuous fizzing and sputtering (AS)

21. The LEAST likely material to be used for a bumper is:
    A. steel.
    B. aluminum.
    C. high-density foam rubber or plastic.
    D. styrofoam. (S)

22. What is the proper grade filter lens when MIG welding?
    A. 3 or 4
    B. 5 or 6
    C. 7 or 8
    D. 9 or 10 (AT)

23. RRIM is a polyurethane composite plastic containing how many parts?
    A. 1
    B. 2
    C. 3
    D. 4 (BO)

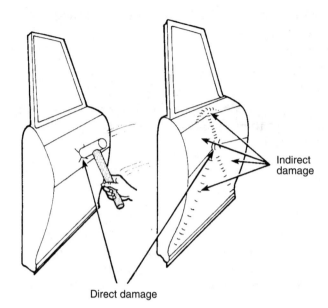

Indirect damage

Direct damage

24. Technician A says direct damage, as shown above, is damage caused by the direct contact of the object causing the damage.
Technician B says indirect damage can be found anywhere on the vehicle.
Who is right?
A. A only
B. B only
C. Both A and B
D. Neither A nor B (L)

25. Technician A says a flash time of 10 minutes is required after applying an adhesive promoter before base color can be applied.
Technician B says newly molded fiberglass parts should be washed with lacquer thinner.
Who is right?
A. A only
B. B only
C. Both A and B
D. Neither A nor B (BP)

26. To align a door properly, it is a good idea to remove the:
A. weather stripping.
B. striker.
C. window glass.
D. hinges. (Q)

27. Technician A says special care should be taken when replacing a hatch because that design is prone to leaking.
Technician B says welded hinges on a hatchback need to be shimmed to make adjustments.
Who is right?
A. A only
B. B only
C. Both A and B
D. Neither A nor B (R)

28. Technician A says on a hatchback, gas-filled shocks are used to adjust alignment. Technician B says a three-way tailgate is difficult to adjust because it has many latches and strikers.
Who is right?
A. A only
B. B only
C. Both A and B
D. Neither A nor B (R)

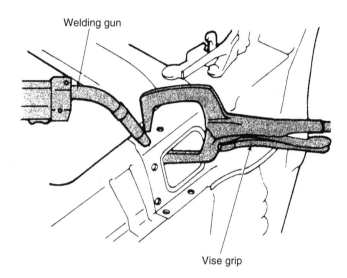

29. When welding, as shown above, all of the following things are used to hold panels tight together at the weld joint EXCEPT:
A. locking pliers.
B. C-clamps.
C. sheet metal screws.
D. fixtures. (AY)

30. When welding two edges of adjacent panels together, it is called a:
A. lap weld.
B. flange weld.
C. plug weld.
D. butt weld. (AZ)

31. When preparing a vehicle for the installation of replacement body panels, technician A grinds the flanges of the structural panels.
Technician B says that weld-through primer must be applied to areas where bare metal is exposed on a joining surface.
Who is right?
A. A only
B. B only
C. Both A and B
D. Neither A nor B (M)

32. Weld-through primer is applied to:
    A. Avoid corrosion starting.
    B. Remove dirt and oil.
    C. Enhance the welding operation.
    D. Strengthen the weld joint. (AY)

33. When removing and installing the hood, you should do all of the following EXCEPT:
    A. Scribe mark the hinges.
    B. Cover and protect adjacent body panels.
    C. Remove the hood hinges.
    D. Align the hood with the cowl and fenders. (O)

34. When adjusting the trunk lid, you must set all of the following EXCEPT:
    A. trunk lid alignment/position.
    B. proper lock assembly engagement.
    C. complete contact with the weather-stripping seal.
    D. torque rod adjustment. (P)

35. Technician A says on a four-door vehicle, it is best to start door adjustment at the front doors and then the rear doors.
    Technician B says worn hinges should be replaced before door adjustments are made.
    Who is right?
    A. A only
    B. B only
    C. Both A and B
    D. Neither A nor B (Q)

36. Technician A says unknown plastic can be identified by international symbols or ISO codes that are molded into plastic parts.
    Technician B says the body repair manual will give information about types of plastics used on the vehicle.
    Who is right?
    A. A only
    B. B only
    C. Both A and B
    D. Neither A nor B (BD)

37. Technician A says hot air plastic welding uses compressed air.
    Technician B says airless welding uses an electric heating element.
    Who is right?
    A. A only
    B. B only
    C. Both A and B
    D. Neither A nor B (BE)

38. What is used to keep dust and water out of the trunk area and to hold the trunk lid tightly shut?
    A. Weather-stripping seal
    B. Latch
    C. Torque rods
    D. Striker (P)

39. What type of weld produces the strongest joint?
    A. V-groove
    B. Lap joint
    C. Flange joint
    D. U-groove (BF)

40. Technician A says solvents should be used to clean the joint before welding. Technician B says a poor weld can be caused by a dirty weld area. Who is right?
    A. A only
    B. B only
    C. Both A and B
    D. Neither A nor B (BF)

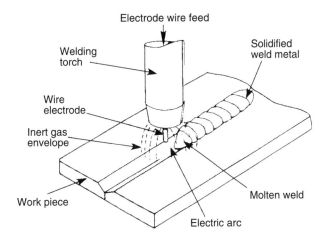

41. When MIG welding, as shown above, what determines proper arc length?
    A. Shielding gas flow
    B. Wire thickness
    C. Penetration
    D. Arc voltage (AS)

42. Technician A says when welding galvanized metals, a respirator should be worn. Technician B says sunglasses are adequate protection from harmful light when MIG welding. Who is right?
    A. A only
    B. B only
    C. Both A and B
    D. Neither A nor B (AT)

43. What is used to help hold the factory panels while the adhesive cures on a sheet molded compound (SMC) vehicle?
    A. Horizontal bracing
    B. Reinforcements
    C. Space-frame
    D. Mill and drill pads (BL)

44. What should be used as a sectioning point when working with an SMC panel?
    A. Panel reinforcements
    B. The space-frame
    C. The mill and drill pads
    D. The horizontal bracing (BL)

45. Old adhesive should be removed from a space-frame using any of the following EXCEPT:
    A. heat.
    B. a putty knife.
    C. a sand blaster.
    D. a sander. (BN)

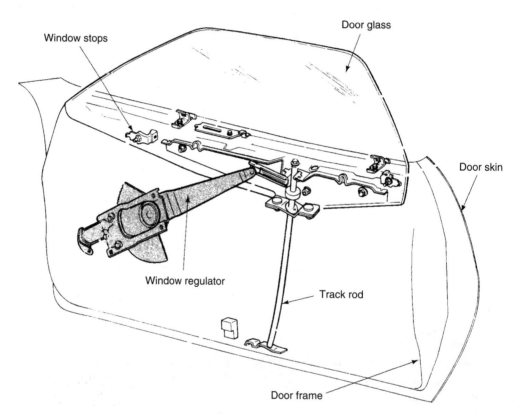

46. As in the figure above, the LEAST likely way of securing door window glass to its frame is using:
    A. bolts.
    B. rivets.
    C. adhesives.
    D. screws. (AK)

47. One of the steps you should perform before welding bare metal is to apply:
    A. weld-through primer.
    B. argon.
    C. flux.
    D. magnesium. (W)

48. When removing or replacing door or quarter glass, it is usually necessary to remove all of the following EXCEPT the:
    A. trim panel.
    B. water shield.
    C. regulator handle.
    D. door skin. (AK)

49. Weather stripping can be held on by all of the following EXCEPT:
    A. glue.
    B. clips.
    C. screws.
    D. a gasket. (AN)

50. Technician A says if different types of materials are being welded, they will not mix correctly, causing a poor or weak quality weld.
    Technician B says a cosmetically poor weld will also be a poor or weak weld.
    Who is right?
    A. A only
    B. B only
    C. Both A and B
    D. Neither A nor B (BB)

51. Cyanoacrylate (CA) is one type of plastic adhesive for automotive plastics. What is the other type?
    A. Resin
    B. Hardener
    C. Super glue
    D. Two-part adhesives (BG)

52. Technician A says some plastics require an adhesion promoter.
    Technician B says to make sure that if any adhesion promoter is used, it dries completely before applying adhesives.
    Who is right?
    A. A only
    B. B only
    C. Both A and B
    D. Neither A nor B (BH)

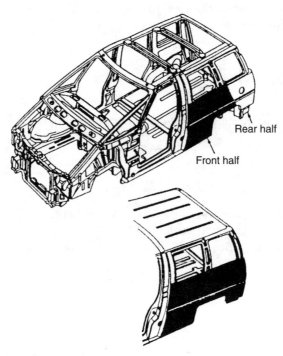

53. Technician A says plastic panel replacement or sectioning, as shown above, will depend on the amount and the location of the damage.
    Technician B says plastic panels cannot be safely sectioned without compromising the vehicle's structural integrity.
    Who is right?
    A. A only
    B. B only
    C. Both A and B
    D. Neither A nor B   (AA)

54. Technician A says brushable seam sealers are used where appearance is important. Technician B says thin bodied sealers should be used to fill seams up to 1/4 inch (6.4 mm) wide.
    Who is right?
    A. A only
    B. B only
    C. Both A and B
    D. Neither A nor B   (AB)

55. What is used to locate the panels in their right location on reinforced plastic panel vehicles?
   A. Locating tabs
   B. Mill and drill pads
   C. Reinforcement panels
   D. Backing strips (AA)

56. Technician A says masking before sanding will protect surfaces not to be painted. Technician B says spray-on masking coatings can be washed off with soap and water.
   Who is right?
   A. A only
   B. B only
   C. Both A and B
   D. Neither A nor B (G)

57. Technician A says all surfaces to be painted should have silicone and wax remover applied before sanding.
   Technician B says even the smallest contamination can ruin a paint job.
   Who is right?
   A. A only
   B. B only
   C. Both A and B
   D. Neither A nor B (H)

58. Technician A says adhesives should be mixed until the color is uniform.
   Technician B says to mix adhesives in the proper proportions according to the product literature.
   Who is right?
   A. A only
   B. B only
   C. Both A and B
   D. Neither A nor B (BG)

59. When repairing a hole in a curved SMC panel, the quickest and cheapest method of repair is the:
   A. molded core.
   B. injection molded.
   C. replacement.
   D. mill and drill. (BM)

60. Technician A says when making a mold, an undamaged panel or vehicle needs to be used as a model or pattern.
   Technician B says a new or used vehicle can be used since it will not be harmed.
   Who is right?
   A. A only
   B. B only
   C. Both A and B
   D. Neither A nor B (BM)

61. Technician A says that it is usually necessary to remove the paint film, undercoat, sealer, or other coatings covering joint areas to find spot welds.
Technician B says heat should not be used to remove coatings that could possibly catch fire.
Who is right?
A. A only
B. B only
C. Both A and B
D. Neither A nor B (I)

62. Technician A says that plastic bumpers should be removed for painting because they are painted with flex agents.
Technician B says plastic parts cannot be repaired and need to be replaced.
Who is right?
A. A only
B. B only
C. Both A and B
D. Neither A nor B (J)

63. If additional adhesive material is needed to fill in a low spot, what should be done?
A. Apply adhesion promoter and apply adhesive.
B. Fill with plastic filler.
C. Sand the area and apply adhesive.
D. V-groove the area and apply adhesive. (BH)

64. Technician A says too much heat can distort a textured surface when welding.
Technician B says a vinyl retextured material can be used to restore the textured surface to the repair area.
Who is right?
A. A only
B. B only
C. Both A and B
D. Neither A nor B (BI)

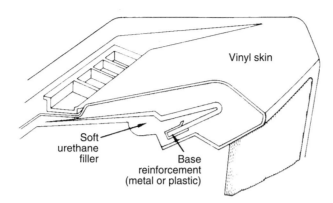

65. Why is vinyl over foam construction, as shown above, used on interior parts?
    A. Cost
    B. Looks
    C. Weight
    D. Safety (BJ)

66. What type of welding relies on vibratory energy to produce a weld?
    A. Ultrasonic
    B. Airless
    C. Hot air
    D. Plasma (BE)

67. Technician A says when a white track is left when the filler is scratched with a fingernail, it is ready to be filed.
    Technician B says a file is used to cut excess filler to size quickly.
    Who is right?
    A. A only
    B. B only
    C. Both A and B
    D. Neither A nor B (AJ)

68. Technician A says that an estimate is also called a damage report.
    Technician B says that a work order outlines the procedure that should be taken to put the vehicle back in preaccident condition.
    Who is right?
    A. A only
    B. B only
    C. Both A and B
    D. Neither A nor B (A)

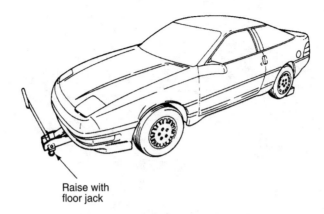
Raise with floor jack

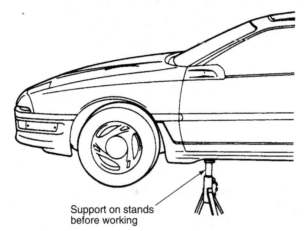
Support on stands before working

69. Which of the following, as shown above, is used to lift the entire automobile?
   A. Floor jack
   B. Hoist
   C. Hydraulic press
   D. Body jack (B)

70. All of the following are methods of heat joining in collision repair EXCEPT:
   A. fusion welding.
   B. pressure welding.
   C. adhesive bonding.
   D. plasma arc welding. (W)

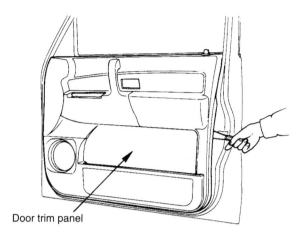
Door trim panel

71. Interior trim panels, as shown above, are fastened by all of the following EXCEPT:
   A. push pins.
   B. snap-in clips.
   C. small screws.
   D. C-clips. (D)

72. An adhesive promoter should always be used on previously painted:
   A. SMC.
   B. RRIM.
   C. fiberglass.
   D. steel. (BO)

73. Technician A says that sometimes you have to remove damaged or undamaged body panels that prevent you from getting to the area of the vehicle being repaired.
   Technician B says it is best to remove bolt-on parts before putting the vehicle on the frame rack so you have better access to the fasteners.
   Who is right?
   A. A only
   B. B only
   C. Both A and B
   D. Neither A nor B (E)

74. Plastic part removal for repair is recommended when:
   A. It is a large part that is difficult to remove.
   B. The repair area is easy to access.
   C. The repair area is small.
   D. The repair area is large. (J)

75. All of the following are potential vehicle safety concerns EXCEPT:
   A. anti-lock brake systems (ABS).
   B. air bag systems (SRS).
   C. emission systems.
   D. steering and suspension systems. (K)

76. Technician A says the repair of aluminum panels requires more care than the repair of steel panels.
    Technician B says aluminum panels are generally thinner than steel panels.
    Who is right?
    A. A only
    B. B only
    C. Both A and B
    D. Neither A nor B (N)

77. Technician A says compressed air is used for both shielding and cutting when plasma arc cutting.
    Technician B says plasma arc cutting can only be used on 1/4 inch (6.4 mm) or thicker metals.
    Who is right?
    A. A only
    B. B only
    C. Both A and B
    D. Neither A nor B (BC)

78. If the back of a replacement hood is not level with the fenders or cowl the technician should:
    A. Adjust the hinges.
    B. Adjust the hood latch.
    C. Adjust the safety catch.
    D. Adjust the bump stops. (O)

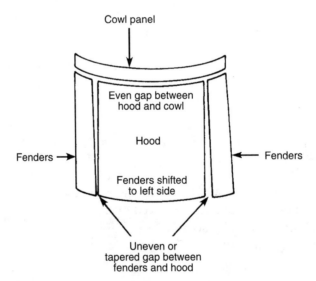

79. Technician A says if the bump stops are not adjusted correctly, as shown above, they may cause the hood to flutter when driving.
    Technician B says the safety catch prevents the hood from flying off should it accidentally open when the vehicle is in motion.
    Who is right?
    A. A only
    B. B only
    C. Both A and B
    D. Neither A nor B (O)

80. Technician A says heat-shrinking is used to work-harden an area of metal. Technician B says you should dolly a damaged area back as close to its original shape as possible before shrinking it.
Who is right?
A. A only
B. B only
C. Both A and B
D. Neither A nor B (AG)

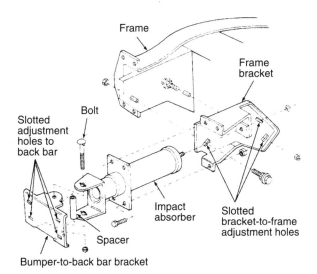

81. Technician A says bumper adjustments, as shown above, are made at the mounting bolts.
Technician B says shims can be added between the bumper and the mounting brackets to adjust bumper alignment.
Who is right?
A. A only
B. B only
C. Both A and B
D. Neither A nor B (S)

82. A replacement fender was just installed on a vehicle. What is normally the proper gap between the fender and the hood/door?
A. 3/16 in (4.8 mm)
B. 1/8 in (3.2 mm)
C. 1/4 in (6.4 mm)
D. 1/16 in (1.6 mm) (T)

83. When a deformed plastic part returns to its original molded shape, it is caused by something called:
A. plastic memory.
B. plasticity.
C. plastic relief.
D. plastic restoration. (BK)

84. Technician A says the edges of mating SMC repair sections should be beveled. Technician B says mill and drill pad nuts should be installed only after all the adhesive has been applied.
    Who is right?
    A. A only
    B. B only
    C. Both A and B
    D. Neither A nor B   (BN)

85. Fenders are bolted down in all of the following areas EXCEPT the:
    A. radiator core support.
    B. inner fender panel.
    C. cowl top, center, and bottom.
    D. valance panel.   (T)

86. Technician A says if the rear of the door can be moved up and down, the hinges are worn and should be replaced.
    Technician B says some hinges use bushings around hinge pins that can be replaced.
    Who is right?
    A. A only
    B. B only
    C. Both A and B
    D. Neither A nor B   (U)

87. The following are all forms of anti-corrosion protection EXCEPT:
    A. seam sealers.
    B. weld-through primers.
    C. rust converters.
    D. two-part electrolytes.   (X)

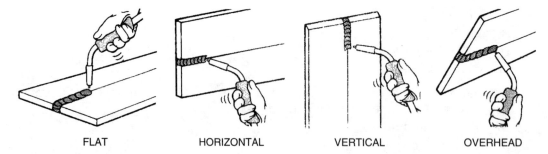

FLAT   HORIZONTAL   VERTICAL   OVERHEAD

88. In what position, as shown above, is it best to keep welding voltage low and weld arc as short as possible?
    A. Overhead
    B. Vertical
    C. Horizontal
    D. Flat   (AW)

89. What type of anti-corrosion protection would you use between two adjacent surfaces?
    A. Seam sealer
    B. Rust converter
    C. Anti-corrosive compound
    D. Two-part electrolyte (X)

90. When replacing a panel at a location other than the factory seams, this process is called:
    A. sectioning.
    B. the joint method.
    C. compound bonding.
    D. structural butt joining. (Y)

91. Technician A says to always follow the procedures described in the manufacturer's body repair manual.
    Technician B says panel replacement performed at the factory seams results in almost identical appearance to factory produced panels.
    Who is right?
    A. A only
    B. B only
    C. Both A and B
    D. Neither A nor B (Y)

92. Technician A says all door skins are welded onto the door frame.
    Technician B says door intrusion beams are normally used in all unibody vehicles.
    Who is right?
    A. A only
    B. B only
    C. Both A and B
    D. Neither A nor B (Z)

93. Foam fillers are used to do all of the following things EXCEPT:
    A. add rigidity and strength.
    B. reduce noise.
    C. reduce vibrations.
    D. give corrosion protection. (AB)

94. The following are all methods of locating air and water leaks EXCEPT:
    A. spraying water on the vehicle.
    B. driving the vehicle over very dirty terrain.
    C. blowing dust over the vehicle.
    D. using a listening device. (AC)

95. Technician A says to wash brazed and soldered joints with soda water to neutralize the acids in the flux.
    Technician B says to wash the repair area with wax and grease remover to eliminate wax, road tar, and grease.
    Who is right?
    A. A only
    B. B only
    C. Both A and B
    D. Neither A nor B   (AF)

96. Technician A says to tape off or remove and store any moldings or nameplates before starting repairs.
    Technician B says undamaged body panels should be removed and stored if they interfere or could be damaged during repairs.
    Who is right?
    A. A only
    B. B only
    C. Both A and B
    D. Neither A nor B   (C)

97. Technician A says when welding horizontally, it is best to angle the gun downward.
    Technician B says when welding overhead, it is best to have a larger puddle.
    Who is right?
    A. A only
    B. B only
    C. Both A and B
    D. Neither A nor B   (AW)

98. What tool(s) can be used to help remove adhesive moldings or nameplates?
    A. Heat gun
    B. Molding tool
    C. Both A and B
    D. Neither A nor B   (C)

99. Technician A says if a brazed joint has been ground down, there is no need to wash the area.
    Technician B says after grinding, blow away the sanding dust with compressed air and wipe the surface with a tack rag to remove remaining dust particles.
    Who is right?
    A. A only
    B. B only
    C. Both A and B
    D. Neither A nor B   (AF)

A. Thin skim coat

B. Fill slightly above panel

100. The first application of filler, as shown above, should be applied:
    A. thin.
    B. thick.
    C. rough.
    D. uncatalyzed. (AI)

101. After grading with a file, you should sand with what grit sandpaper?
    A. 36 grit paper
    B. 80 grit paper
    C. 120 grit paper
    D. 150 grit paper (AJ)

102. Technician A says you should disconnect the battery before welding to protect electronics.
    Technician B says connect the ground as close as possible to the work to avoid a current seeking its own ground and damaging the vehicle.
    Who is right?
    A. A only
    B. B only
    C. Both A and B
    D. Neither A nor B (AX)

103. Technician A says any scratches made by the file should be removed with sandpaper.
    Technician B says oversanding will make it necessary to apply more filler.
    Who is right?
    A. A only
    B. B only
    C. Both A and B
    D. Neither A nor B (AJ)

104. A guide coat is applied and sanded, and some areas of the primer mist are left intact. This means there is a:
    A. low spot.
    B. high spot.
    C. soft spot.
    D. wet spot. (AJ)

105. Technician A says when installing weather stripping to stretch it tight.
    Technician B says a sponge rubber plug is often used to hold the cut ends of weather stripping together.
    Who is right?
    A. A only
    B. B only
    C. Both A and B
    D. Neither A nor B (AN)

106. Technician A says interior trim panels should be covered if repairs could damage them.
    Technician B says any parts interfering with repairs should be removed.
    Who is right?
    A. A only
    B. B only
    C. Both A and B
    D. Neither A nor B (D)

107. Technician A says convertible tops are not adjustable.
    Technician B says convertible tops should be checked for leaks and wind noise after a collision.
    Who is right?
    A. A only
    B. B only
    C. Both A and B
    D. Neither A nor B (AO)

108. What determines the welding process used to make the repair?
    A. Manufacturers' recommendations
    B. Shop procedures
    C. Tools available
    D. Technician's welding experience (AR)

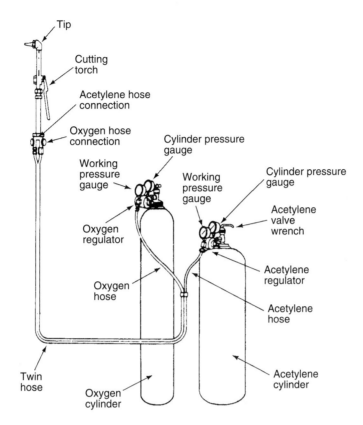

109. Technician A says gas cylinders, as shown above, are made to take the abuse of a shop environment.
Technician B says gas cylinders might be pressurized to more than 2,000 psi (13,790 kPa).
Who is right?
   A. A only
   B. B only
   C. Both A and B
   D. Neither A nor B (AU)

110. Technician A says the welding machine's cable clamp is not really a grounding cable.
Technician B says the ground connection is used for safety purposes only.
Who is right?
   A. A only
   B. B only
   C. Both A and B
   D. Neither A nor B (AV)

111. What position is the easiest to weld in?
   A. Overhead
   B. Vertical
   C. Horizontal
   D. Flat (AW)

112. Today's high-strength, low-alloy steel can only be welded properly using:
    A. arc welding.
    B. MIG welding.
    C. soldering.
    D. brazing. (AQ)

113. Technician A says to ground yourself before handling electronic equipment to avoid static electricity and damage to the part.
    Technician B says to cover any computer that may be affected by work being performed.
    Who is right?
    A. A only
    B. B only
    C. Both A and B
    D. Neither A nor B (AX)

114. All of the following are welding problems EXCEPT:
    A. weld porosity.
    B. weld spatter.
    C. weld dents.
    D. weld cracks. (BB)

115. When electronic parts and their connectors are removed they should be protected by putting them in:
    A. the trunk of the vehicle.
    B. plastic cases.
    C. a wooden box.
    D. anti-static bags. (AX)

116. What type of metal cannot be welded?
    A. Steel
    B. Aluminum
    C. Stainless steel
    D. Magnesium (AP)

117. Technician A says avoid running welding cables close to electronic displays or computers.
    Technician B says avoid touching bare electrical contacts to keep from getting a shock.
    Who is right?
    A. A only
    B. B only
    C. Both A and B
    D. Neither A nor B (AX)

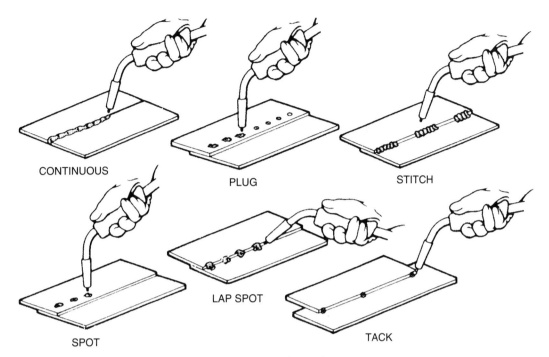

118. A plug weld, as shown above, is a body shop alternative to a:
    A. spot weld.
    B. butt weld.
    C. lap weld.
    D. flange weld. (AZ)

119. Technician A says lap and flange welds should be used on structural panels. Technician B says when warpage is a concern, you should perform a butt weld. Who is right?
    A. A only
    B. B only
    C. Both A and B
    D. Neither A nor B (AZ)

120. Technician A says a magnet can be used to find out if a metal is steel. Technician B says if a metal is not magnetic and it turns dull gray when brushed with a stainless steel brush, it is aluminum. Who is right?
    A. A only
    B. B only
    C. Both A and B
    D. Neither A nor B (AP)

121. Stitch welding combines continuous welding with:
    A. tack welding.
    B. MIG welding.
    C. pulse welding.
    D. spot welding. (BA)

122. Technician A says dash pads are usually replaced when damaged.
    Technician B says dash pads are expensive and time-consuming to repair.
    Who is right?
    A. A only
    B. B only
    C. Both A and B
    D. Neither A nor B (BJ)

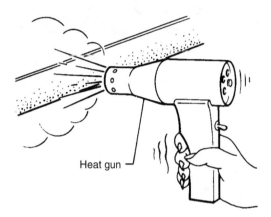

123. Technician A says deformed plastic parts, as shown above, can often be straightened with heat.
    Technician B says very few types of plastic parts can be straightened with heat.
    Who is right?
    A. A only
    B. B only
    C. Both A and B
    D. Neither A nor B (BK)

124. Technician A says heat should NOT be applied to RRIM to help curing because it is highly flammable.
    Technician B says a backing patch is required if damage or a puncture extends through the RRIM panel.
    Who is right?
    A. A only
    B. B only
    C. Both A and B
    D. Neither A nor B (BO)

6 Additional Test Questions for Practice

Additional Test Questions

Please note the letters in parentheses following each sample question. These letters match the overview in section 4 that discusses the relevant subject matter. You may want to refer back to the overview using this cross-referencing key to help with questions posing problems for you.

1. All of the following are recommended to remove paint film or other coatings covering joints EXCEPT:
    A. DA sander.
    B. wire wheel or brush.
    C. scraper.
    D. oxyacetylene or propane torch. (I)

2. Technician A says welded hinges on a hatch are not adjustable.
    Technician B says to refer to the body repair manual to find out how the hatch is adjusted.
    Who is right?
    A. A only
    B. B only
    C. Both A and B
    D. Neither A nor B (R)

3. The paint system for flexible plastic parts might require the addition of a:
    A. hardener.
    B. fisheye preventer.
    C. flex agent.
    D. adhesive promoter. (J)

4. Technician A says entrance dust is usually found in areas with an air or water leak.
    Technician B says these points should be sealed with appropriate sealing compounds and then checked to verify that the leak is sealed.
    Who is right?
    A. A only
    B. B only
    C. Both A and B
    D. Neither A nor B (AC)

5. What type of plastic welding uses compressed air?
   A. Ultrasonic welding
   B. Hot air plastic welding
   C. Forced hot air welding
   D. Airless plastic welding (BE)

6. Technician A says worn parts undamaged from the accident should be reported to the customer for additional repairs.
   Technician B says vehicle safety items should be closely checked before returning the vehicle to the customer.
   Who is right?
   A. A only
   B. B only
   C. Both A and B
   D. Neither A nor B (K)

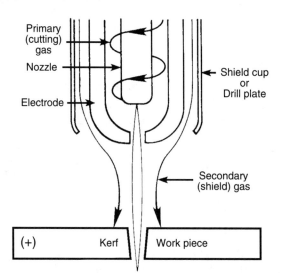

7. What type of process, as shown above, creates an intensely hot air stream over a very small area that melts and removes metal?
   A. Arc cutting
   B. Oxyacetylene torch cutting
   C. Laser cutting
   D. Plasma arc cutting (BC)

8. Good welding practices require clamping panels together for sound welds. All of the following are used for that purpose EXCEPT:
   A. locking pliers or clamps.
   B. sheet metal screws.
   C. tack welds.
   D. weld-through primer. (AV)

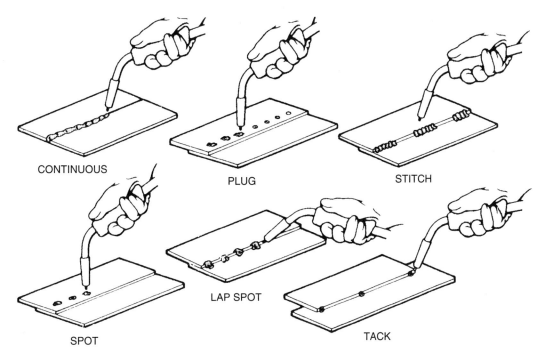

9. What type of weld, as shown above, can be used anywhere in the body that the factory used a resistant spot weld?
   A. Plug weld
   B. Lap weld
   C. Flange weld
   D. Butt weld (AZ)

10. Technician A says a sunroof that leaks can be out of adjustment or misaligned. Technician B says to consult the manufacturer's service manual for adjustment procedures.
    Who is right?
    A. A only
    B. B only
    C. Both A and B
    D. Neither A nor B (AL)

11. The most important point to consider when sectioning is:
    A. how it will affect the estimate.
    B. how long it will take.
    C. how it will affect structural integrity.
    D. what the customer prefers. (Y)

12. Exterior trim may get damaged during repairs if not removed.
    Technician A says if the trim has to be removed it will be replaced.
    Technician B says some trim may require special tools for removal.
    Who is right?
    A. A only
    B. B only
    C. Both A and B
    D. Neither A nor B (C)

13. Heat can be used to repair flexible plastic parts. All of the following types of damage can be repaired this way EXCEPT:
    A. rips.
    B. stretches.
    C. deformations.
    D. bends. (BK)

14. Applying mixed filler thickly without a thin coat first causes:
    A. poor bonding and pinholes.
    B. pinholes and wrong hardening.
    C. cracking and poor bonding.
    D. rough surface and poor bonding. (AI)

15. Technician A says the first thing you should do when removing a hood is scribe mark the hood latch.
    Technician B says you should have a second technician help you install and align the hood.
    Who is right?
    A. A only
    B. B only
    C. Both A and B
    D. Neither A nor B (O)

16. The trunk lid is similar in construction to the:
    A. doors.
    B. fenders.
    C. hood.
    D. rear body panel. (P)

17. The molded core method is generally used to repair a curved SMC panel with:
    A. a crack or tear.
    B. holes.
    C. dents.
    D. spider webbing. (BM)

18. Where can a technician find out what type of plastic is used on a vehicle?
    A. Service manual
    B. Body repair manual
    C. Owner's manual
    D. Estimate (BD)

19. What outlines the procedure to be taken to put the vehicle back in preaccident condition?
   A. Work order
   B. Estimate
   C. Damage report
   D. Appraisal (A)

20. RRIM is a two-part polyurethane composite plastic. All of the following materials are used in Part B of the mixture EXCEPT:
   A. isocyanate.
   B. reinforcing fibers.
   C. resins.
   D. a catalyst. (BO)

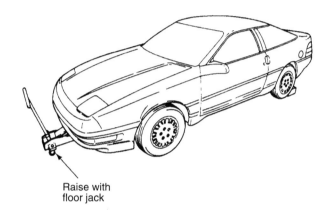
Raise with floor jack

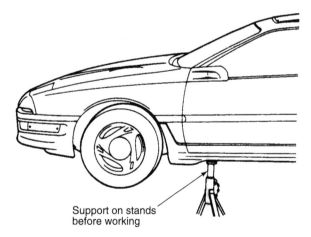
Support on stands before working

21. Where can a technician find the right lifting and jacking points, as shown above, for a vehicle?
   A. Jacking label
   B. Hoist label
   C. Service manual
   D. Body repair manual (B)

22. What should newly molded parts be washed with before painting or priming?
    A. Adhesive promoter
    B. Lacquer thinner
    C. Denatured alcohol
    D. Turpentine (BP)

23. Where can a technician find the locations for the fasteners holding interior trim parts?
    A. Owner's guide
    B. Technician's guide
    C. Service manual
    D. Body repair manual (D)

24. What type of weld requires special welding nozzles and settings to be used on a MIG welder?
    A. Plug weld
    B. Lap weld
    C. Flange weld
    D. Spot weld (BA)

25. Deciding when it is convenient to remove bolt-on parts before making repairs is dependent on all of the following EXCEPT:
    A. the vehicle.
    B. the location of damage.
    C. the degree of damage.
    D. the estimate. (E)

26. Technician A says aluminum body tape can be used to align the break on plastic parts before welding.
    Technician B says a razor blade can be used to shape excess weld buildup and smooth contour the weld.
    Who is right?
    A. A only
    B. B only
    C. Both A and B
    D. Neither A nor B (BF)

27. When welding overhead, as shown above, it is best to keep the weld puddle:
    A. small.
    B. big.
    C. hot.
    D. cool. (AW)

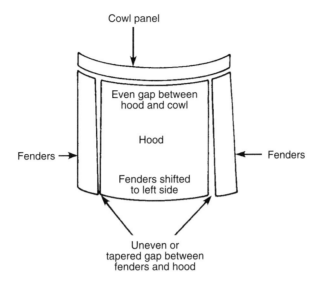

28. Fenders, as shown above, should be adjusted/aligned after:
    A. hood adjustment.
    B. bumper adjustments.
    C. header panel adjustments.
    D. door adjustments. (T)

29. Technician A says to assume that contaminants are present and to always include some wax and grease cleaner or detergent in the sanding water.
    Technician B says wax and silicone do not penetrate the outer surface and can be sanded out.
    Who is right?
    A. A only
    B. B only
    C. Both A and B
    D. Neither A nor B (H)

30. The best way to monitor heat application when stress relieving is to use:
    A. a heat crayon.
    B. a temperature gauge.
    C. the sheet metal color.
    D. your hand. (L)

31. Gas cylinders either on a cart or in the shop should always be:
    A. filled.
    B. emptied.
    C. heated.
    D. supported. (AU)

32. Aluminum has all of the following qualities EXCEPT:
    A. it is softer than steel.
    B. it is heavier than steel.
    C. it melts at a lower temperature.
    D. it is stiffer to the touch than steel. (N)

33. On a four-door sedan door adjustment, you should always start adjustments with the:
    A. top hinges.
    B. bottom hinges.
    C. front doors.
    D. rear doors. (Q)

34. Welding at too rapid a pace can cause all the following EXCEPT:
    A. poor penetration.
    B. thin bead width.
    C. dome shaped bead.
    D. burn-through holes. (AS)

35. On older cars, bumpers were rigidly bolted to:
    A. the fenders.
    B. the core support.
    C. the frame.
    D. the energy absorbers. (S)

36. When welding, clamping both sides of the panel is not always possible. In these cases, all of the following can be used EXCEPT:
    A. self-tapping screws.
    B. rivets.
    C. fixtures and clamps.
    D. adhesives. (W)

37. What is used to cut excess filler to size quickly?
    A. A DA sander
    B. A grinder
    C. A file
    D. Sandpaper (AJ)

38. What parts are foam fillers generally used in?
    A. Nonstructural sheet metal
    B. Doors
    C. Structural pillars/rockers
    D. Frame rails (AB)

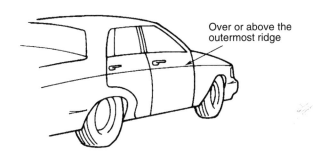

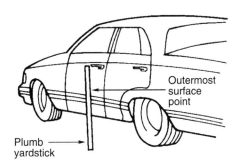

39. Where should aftermarket body side moldings, as shown above, be placed?
    A. The outermost surface of the vehicle
    B. The center of the door
    C. On a body line
    D. The same place as the factory molding (AD)

40. Technician A says if a part is difficult to mask, it should be removed.
    Technician B says if a part can be removed easily, it should be.
    Who is right?
    A. A only
    B. B only
    C. Both A and B
    D. Neither A nor B (G)

41. Technician A says if the door glass has to be aligned with the edge of the quarter glass to make sure the quarter glass is adjusted properly first.
    Technician B says the manufacturer's procedures should be consulted when adjusting windows.
    Who is right?
    A. A only
    B. B only
    C. Both A and B
    D. Neither A nor B (AK)

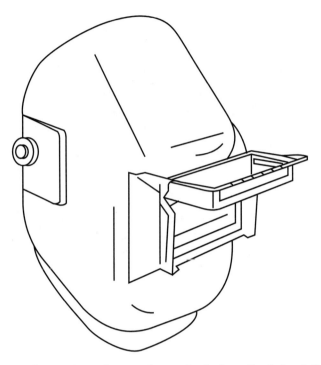

42. Welding safety equipment, as shown above, includes all of the following EXCEPT:
    A. Gloves
    B. Glasses
    C. Respirators
    D. Boots (AT)

43. Removable manually operated roof panels should be inspected for all the following EXCEPT:
    A. cracks.
    B. leaks.
    C. broken hardware.
    D. warpage. (AM)

44. Technician A says to use high heat when working with textured vinyl. Technician B says textured vinyl can be retextured.
    Who is right?
    A. A only
    B. B only
    C. Both A and B
    D. Neither A nor B (BI)

Additional Test Questions for Practice　　　　　　　　　　　Additional Test Questions　97

45. Technician A says to carefully study the locations of the engine mounts and suspension mounts, and whether or not these parts themselves are damaged.
    Technician B says you cannot straighten a vehicle with the suspension and driveline installed.
    Who is right?
    A. A only
    B. B only
    C. Both A and B
    D. Neither A nor B　　　　　　　　　　　　　　　　　　　　　　　　　　(F)

46. Weather stripping usually fits over:
    A. a drip rail.
    B. a pinch weld flange.
    C. screws.
    D. clips.　　　　　　　　　　　　　　　　　　　　　　　　　　　　　　(AN)

47. Technician A says if weld-through primer is applied, no other corrosion protection is needed.
    Technician B says seam sealers are only used on areas that might leak water into the interior of the car.
    Who is right?
    A. A only
    B. B only
    C. Both A and B
    D. Neither A nor B　　　　　　　　　　　　　　　　　　　　　　　　　　(X)

48. Technician A says new panels can be visually positioned by noting the relationship between the surrounding panels and the new part when repairing minor damage.
    Technician B says that dimension measuring instruments should be used to determine correct part positions of new panels when repairing major damage.
    Who is right?
    A. A only
    B. B only
    C. Both A and B
    D. Neither A nor B　　　　　　　　　　　　　　　　　　　　　　　　　　(AY)

49. What type of welding technique has replaced earlier techniques and become commonplace?
    A. Arc welding
    B. Oxyacetylene welding
    C. MIG welding
    D. Spot welding　　　　　　　　　　　　　　　　　　　　　　　　　　　(AQ)

50. Too much weld penetration will cause the weld to:
    A. be even with the base metal.
    B. sit on top of the base metal.
    C. burn through the base metal.
    D. be uneven and warp the base metal.　　　　　　　　　　　　　　　　　(BB)

51. Plastic part cracks and rips can be repaired in two ways, one is welding. What is the other?
    A. Plastic fillers
    B. Fiberglass fillers
    C. Adhesives
    D. Resins (BG)

52. Door skins are secured to the door frame by:
    A. screws or bolts.
    B. welding or adhesives.
    C. the flange around the frame.
    D. plastic fillers. (Z)

53. Vehicles using reinforced plastic panels are supported by a:
    A. space-frame.
    B. fiberglass matte.
    C. full frame.
    D. tubular structure. (AA)

54. When making repairs, an adhesion promoter is used on:
    A. all plastics.
    B. most plastics.
    C. some plastics.
    D. damaged plastics. (BH)

55. A technician cannot get enough hinge adjustment to align the door properly. Technician A says to replace the door hinges because they are probably worn. Technician B says to see if a hinge pin and bushing kit is available to repair the hinges.
    Who is right?
    A. A only
    B. B only
    C. Both A and B
    D. Neither A nor B (U)

56. On sheet metal, a straight file is used to:
    A. flatten the metal.
    B. find high and low spots.
    C. prepare the area for filler.
    D. sand the area. (V)

57. Shrinking metal uses heat usually from an oxyacetylene torch with:
    A. a #1 or #2 tip.
    B. a #2 or #3 tip.
    C. a #3 or #4 tip.
    D. a #4 or #5 tip. (AG)

58. Technician A says wire shielding can be left out if damaged from a wiring harness. Technician B says electronic crossover from current-carrying wires can affect sensing and control circuits.
    Who is right?
    A. A only
    B. B only
    C. Both A and B
    D. Neither A nor B   (AX)

59. A mixing board is usually made of all of the following EXCEPT:
    A. metal.
    B. glass.
    C. plastic.
    D. cardboard.   (AH)

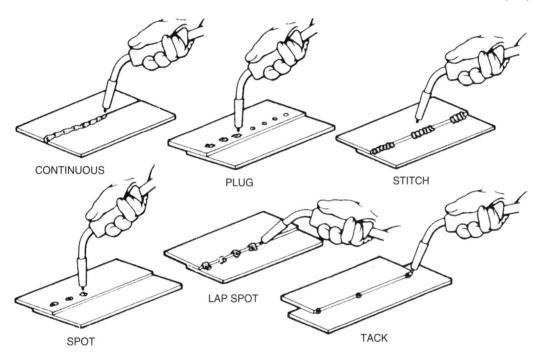

60. What type of weld, as shown above, is removed using a drill?
    A. Spot
    B. Butt
    C. Lap
    D. Stitch   (M)

61. Vinyl over a foam filler is commonly used in all of the following EXCEPT:
    A. vinyl tops.
    B. instrument panels.
    C. seats.
    D. headliners   (BJ)

Page 101, answers should read: 8. D, 15. C, 21. D, 45. C, 58. C, 80. B, 84. A, 88. A, 101. A, and 107. B
Page 102, answers should read: 1. D, 3. C, 5. B, 6. C, 7. D, 8. D, 9. A, 12. B, 14. A, 17. B, 18. B, 19. A, 20. A, 22. C, 25. D, 26. C, 28. A, 29. A, 31. D, 32. B, 34. D, 36. D, 37. C, 38. C, 39. D, 40. C, 41. C, 44. B, 45. A, 46. B, 47. D, 48. C, 50. C, 51. C, 53. A, 57. A, 58. B, 59. D, and 61. D

Appendices

Answers to the Test Questions for the Sample Test Section 5

1.	A	32.	A	63.	A	94.	C
2.	B	33.	C	64.	C	95.	C
3.	B	34.	D	65.	D	96.	C
4.	D	35.	B	66.	A	97.	D
5.	A	36.	C	67.	C	98.	C
6.	B	37.	C	68.	C	99.	B
7.	C	38.	A	69.	B	100.	A
8.	A	39.	A	70.	D	101.	B
9.	D	40.	B	71.	D	102.	C
10.	B	41.	D	72.	A	103.	C
11.	B	42.	A	73.	C	104.	A
12.	B	43.	D	74.	A	105.	B
13.	A	44.	A	75.	C	106.	B
14.	B	45.	D	76.	A	107.	C
15.	A	46.	D	77.	A	108.	A
16.	B	47.	A	78.	A	109.	B
17.	C	48.	D	79.	C	110.	C
18.	D	49.	D	80.	C	111.	D
19.	C	50.	C	81.	C	112.	B
20.	A	51.	D	82.	A	113.	A
21.	B	52.	C	83.	A	114.	C
22.	D	53.	A	84.	C	115.	D
23.	B	54.	D	85.	D	116.	D
24.	C	55.	B	86.	C	117.	A
25.	D	56.	C	87.	D	118.	A
26.	B	57.	C	88.	D	119.	A
27.	A	58.	B	89.	A	120.	A
28.	B	59.	A	90.	A	121.	D
29.	D	60.	C	91.	C	122.	D
30.	D	61.	A	92.	B	123.	A
31.	C	62.	A	93.	D	124.	B

Answers to the Test Questions for the Additional Test Questions Section 6

1.	A	17.	D	33.	A	49.	C
2.	C	18.	C	34.	C	50.	D
3.	B	19.	C	35.	C	51.	B
4.	C	20.	D	36.	C	52.	B
5.	D	21.	C	37.	D	53.	C
6.	A	22.	B	38.	B	54.	C
7.	C	23.	D	39.	C	55.	C
8.	A	24.	D	40.	D	56.	B
9.	D	25.	C	41.	D	57.	D
10.	C	26.	B	42.	D	58.	A
11.	C	27.	A	43.	D	59.	B
12.	A	28.	C	44.	A	60.	A
13.	A	29.	C	45.	B	61.	C
14.	B	30.	A	46.	C		
15.	B	31.	A	47.	A		
16.	C	32.	D	48.	D		

Glossary

A Abbreviation for ampere.

Abrasive A material such as sand, crushed steel grit aluminum oxide, silicon carbide, or crushed slag used for cleaning or surface roughening.

Abrasive coating (1) In closed coating paper, the complete surface of the paper is covered with abrasive; no adhesive is exposed. (2) In open coating, adhesive is exposed between the grains of abrasive.

Accent stripes Lines applied to a vehicle to add a decorative, customized look.

Access hole An opening that permits a technician to access fasteners and other components inside a door.

Access time The time required to remove extensively damaged collision parts by cutting, pushing, or pulling.

Accessible area An area that can be reached without parts being removed from the vehicle.

Accessories Items that are not essential to the operation of a vehicle, such as the cigar lighter, radio, luggage rack, or heater.

Acetylene A gas used in flame welding and cutting.

Acid core A type of solder in tubular wire form having an acid flux paste core.

Acrylic A thermoplastic synthetic resin used in both emulsion and solvent-based paints, available as a lacquer or enamel.

Acrylic enamel A type of finish that contains polyurethane and acrylic additives.

Acrylic polyurethane enamel A material with great weatherability that generally provides higher gloss and greater durability than other polyurethane enamels.

Activator An additive used to cure a two- or multi-package enamel.

Active restraint A seat belt that the occupant of a vehicle must fasten.

Actual cash value (ACV) The current market value of a standard production vehicle and its accessory options as determined by used car guidebook listings or car dealer assessments.

Additive A chemical substance that is added to a finish, in small amounts, to impart or improve desirable properties.

Adhesion The ability of one substance to stick to another.

Adhesion promoter A water-white, ready-to-spray lacquer material that provides a chemical etch to original equipment manufacturer (OEM) finishes.

Adhesive backed molding A trim piece provided with an adhesive back coating to simplify installation.

Adhesive bonding A mechanical bonding between the adhesive and the surfaces that are being joined together.

Adhesive caulk A material used to seal or join seams, and used to install windshields and rear window glass.

Adhesive compound A non-hardening caulk-type material used to hold fixed glass in place.

Adhesive joining The assembly of components with a chemical bonding agent.

Adhesive primer/hardener A material brushed on a mirror support and on the glass before applying the mirror mounting adhesive.

Adjuster An insurance company representative, often called an appraiser, responsible for approving a collision repair bid to satisfy a vehicle damage claim.

Adjusting slot The elongated holes on mounting brackets and bumper shock absorbers that permit alignment during installation.

Adjustment The bringing of parts into alignment or proper dimensions with fasteners that permit some movement for position.

Aerodynamic A body shape having a low wind resistance.

Aftermarket Any parts or accessories, new or used, that are installed after original manufacture of the vehicle.

Aging The process of permitting a material to stand for a period of time.

Agitator A paint mixer or stirrer of any type.

Aiming screw The adjusting screws used to aim a headlight.

Air (1) A natural gas, usually used under pressure as a propellant. (2) The natural gas we breathe.

Air bag system A system that uses impact sensors, an on-board computer, an inflation module, and a nylon bag in the dash and/or steering column to protect the passenger and/or driver during a collision.

Air brush A small paint spray device used for fine detailing, fish scaling, and similar paint work.

Air cap The component located at the front of the gun that directs the compressed air into the material stream.

Air compressor Equipment that is used to supply pressurized air to operate shop tools and equipment.

Air conditioning system A system having a compressor, evaporator, condenser, and associated components that cool the air in the passenger compartment of a vehicle.

Air dam The structure mounted under the front portion of a vehicle designed to direct air through the radiator and across the engine.

Air drying The act of allowing a painted surface to dry at ambient temperature without the aid of an external source.

Air filter A device used to trap dirt particles or other debris in an air line.

Air gun A device that uses compressed air to clean and dry surfaces and areas.

Air line A pipe or flexible hose used to transport compressed air from one point to another.

Air make-up system A method used to replace the air that is exhausted from a paint spray booth.

Air-over-hydraulic system A system that utilizes a pneumatic motor to drive a hydraulic pump.

Air pressure (1) The force exerted on a container by compressed air. (2) The pressure of the surrounding ambient air.

Air-purifying respirator A filtered breathing aid used to clean or purify air.

Air spray A method of applying paint in the form of tiny droplets in air as paint is atomized by a spray gun as a result of being forced into a high velocity air stream.

Air supply system An air pumping system to supply fresh air for breathing to a painter in a paint spray booth.

Air suspension A vehicle suspension system that makes use of pneumatic cylinders to replace or supplement mechanical springs.

Air transformer A pneumatic control device that is used to filter and control the air delivered by a compressor.

Airless spraying A method of spray paint application in which atomization is effected by forcing paint, under high pressure, through a very small orifice in a spray gun cap.

Alcohol A colorless volatile liquid (1) used as diluents, solvents, or co-solvents in paints and (2) used as a fuel for racing engines.

Align To make an adjustment to a line or to a predetermined relative position.

Alignment The arrangement of a vehicle's basic structural components in relation to each other.

Alignment gap The space between two components, such as a door and pillar or fender and hood.

Alkyd A chemical combination of an alcohol, an acid, and an oil useful in water-based house paint and automotive primers.

Alkyd enamel The least expensive of the enamels.

Alligatoring A paint finish defect that resembles the pattern of an alligator's skin.

Alloy A mixture of two or more metals.

Alternating current Electrical current that changes direction.

Alternator An electrical device that is used to generate alternating current, which is then rectified into direct current for use in the vehicle's electrical system.

Aluminum (Al) (1) A lightweight metal. (2) A material that is useful as a substrate or pigment.

Aluminum oxide An extremely tough abrasive that is highly resistant to fracturing and capable of penetrating hard surfaces without dulling.

Ambient temperature The temperature of the surrounding air.

Ammeter An electrical instrument that is used to measure electrical current.

Amp An abbreviation for ampere.

Ampere (A) An electrical unit for current.

Anchor To hold in place.

Anodizing An electrolytic surface treatment for aluminum that builds up an aluminum oxide coating.

Antenna wire circuit A circuit placed between layers of glass or printed on the interior surface of the glass.

Anti-fouling A paint containing toxic substances that inhibit the growth of certain organisms on ship bottoms.

Anti-lacerative glass A form of laminated glass having an additional layer of plastic on the occupant side to contain fragments of glass during breakage.

A-pillar The windshield post.

Appraiser An insurance company representative, often referred to as an adjuster, responsible for approving a collision repair bid.

Apprentice A person who learns his or her trade while working under the supervision of a skilled technician concurrent with classroom training.

Arc spot welding The process of making spot welds with the heat of an arc, primarily in areas that are not accessible with resistance spot welders.

Arc welding A joining technique that uses an electrical arc as the heat source.

Asbestos A cancer-causing material that was once used in the manufacture of brake and clutch linings.

Asbestos dust A cancer-causing byproduct agent of a material that was once used in the manufacture of brake and clutch lining assemblies.

ASE certification A voluntary testing program to help to give recognition that you are a knowledgeable collision repair technician, estimator, or painter.

Asphyxiation The inability to breathe due to anything that prevents normal breathing, such as mists, gases, and fumes.

Assembly Two or more parts that are bolted or welded together to form a single unit.

Assembly drawing Drawings used by the manufacturer when assembling a component or a vehicle.

Atmosphere-supplied respirator A system that has a blower or a special compressor to supply clean fresh outside air to a face mask or hood.

Atomize To break a liquid into a fine mist.

Auto-ignition temperature The approximate lowest temperature at which a flammable gas or vapor-air mixture will ignite without spark or flame.

Automatic cut-off A safety device used to shut off the air compressor at a preset pressure.

Automatic welding Welding equipment that performs the welding operation without constant observation and adjustment of the controls.

Average retail price The local value of a like vehicle, based on actual sales reports from new and used car dealers.

Back light A vehicle window located behind the occupants.

Backfire (1) A small explosion that can cause a sharp popping sound in oxygen-acetylene welding equipment. (2) An explosion in the exhaust system of a motor vehicle. (3) An explosion in the intake manifold of a vehicle.

Baffle A panel used to direct air to the radiator.

Baked enamel A finish achieved when heat is used to promote rapid drying.

Baked-on finish A painted surface that has been cured by heating after application.

Banding A single coat applied in a small spray pattern to frame in an area to be painted.

Bar gauge A gauge used to accurately measure and diagnose body and frame collision damages for all conventional and unitized vehicles.

Appendices
Glossary

Barrier cream A hand cream that provides protection and soothes the skin when working with irritating materials.

Base The resin component of paint to which color pigments and other components are added.

Basecoat The coat of paint on which final coats are applied.

Basecoat/clearcoat A paint method in which the color effect is given by a highly pigmented basecoat followed by a clearcoat to provide gloss and durability.

Base metal Any metal to be welded or cut.

Bathtub clip Small plastic attachment pieces that are pressed into holes in the vehicle body to which moldings are attached.

Battery charger An electrical device used to recharge a vehicle battery.

Bead The amount of filler metal or plastic material deposited in a joint when welding two pieces together.

Bearing plate A plastic spacer used to reduce the friction between the door handle and the trim panel.

Belt line A horizontal molding or crown along the side of the vehicle at the bottom of the glass.

Belt trim support retainers The parts that hold the window glass in position and prevent it from wobbling.

Bench A vehicle underbody anchoring device for checking frame and suspension dimensions for damage and to allow straightening procedures.

Bench grinder A tool used for sharpening or metal removal that is bolted to a workbench and driven by an electric motor.

Bench system An alignment method using equipment that allows the vehicle to be set on preadjusted pins to check for damage.

Bench vise A heavy adjustable bench mounted holding tool.

Bend A change in a body or trim part from its original shape.

Bezel A trim ring that surrounds a headlight or gauge.

Binder An ingredient in paint that holds pigment particles together.

Bleeder gun A spray gun design that allows air to pass through the gun at all times, preventing pressure buildup in the supply lines.

Bleeding The original color showing through after a new coat has been applied.

Blistering The formation of hollow bubbles or water droplets in a paint film, usually caused by expansion of air or moisture trapped beneath the paint film.

Blending (1) The mixing of two or more paint colors to achieve the desired color. (2) The technique used in repairing acrylic lacquer finishes, extending each color coat a little beyond the previous coat to blend into surrounding finish.

Blistering The formation of bubbles on the surface of paint, usually caused by moisture or dirt.

Block sanding The use of a flat object, such as a block of wood, and sandpaper to obtain a flat surface.

Bloom A clouded appearance on a finish paint coat.

Blowgun A device that uses blasts of compressed air to help clean and dry work surfaces.

Blushing The hazing or whitening of a film caused by absorption and retention of moisture in the drying paint film.

Bodied cement A syrupy solvent cement that is composed of a solvent and a small quantity of compatible plastic.

Body (1) The consistency of a liquid; the apparent viscosity of a paint as assessed when stirring it. (2) An assembly of sheet metal parts that comprise the enclosure of a vehicle.

Body code plate A manufacturer-mounted plate on a vehicle providing body type and other information.

Body file A flat, half-round, or round file designed to be pushed across the work surface in the direction of the cutting teeth.

Body filer A heavy-bodied plastic-like material that cures very hard; used to fill small dents in metal.

Body hammer A specialized hammer used to reshape damaged sheet metal.

Body hardware Appearance and functional parts, such as handles, on the interior and exterior of a vehicle.

Body mounting The method and means by which a vehicle body is placed onto a chassis.

Body-over-frame A vehicle having separate body and chassis parts bolted to the frame.

Body-over-frame construction A method in which the automobile body is bolted to a separate frame.

Body panel A sheet of shaped steel, aluminum, or plastic that forms part of the car body.

Body saw A saw equipped with an abrasive blade used to cut floor sections or panels.

Body solder An alloy of tin and lead, used to fill dents and other body defects.

Body technician A person who performs such basic repair tasks as removing dents, replacing damaged parts, welding metal, filling, sectioning, and sanding.

Body trim (1) The material used to finish the interior of the passenger and trunk compartments. (2) The rubber and metal moldings on the exterior of a vehicle.

Bolt-through glass Glass that is held to the regulator mechanism with mechanical fasteners, such as bolts or rivets.

Bonding strips Strips of aluminum, fiberglass, or aluminum and stainless steel tape, used to patch holes in vehicle bodies.

Bounce-back The atomized particles of paint that rebound from the surface being sprayed and contribute to overspray.

B-pillar The pillar between the belt line and roof between the front and rear doors on four-door vehicles and station wagons.

Brackets A part that is used to attach components to each other or to the body and frame.

Braze welding A method of welding using a filler metal.

Bridging The characteristic of an undercoat that occurs when a scratch or surface imperfection is not completely filled.

Brittle A lack of toughness and flexibility.

Bronzing The formation of a metallic-appearing haze on a paint film.

Brush (1) A method of applying paint. (2) An applicator for applying paint.

Brushing The act of applying paint using a brush.

Buckles The distortion, ridges, or high places on a metal body part as a result of collision damage.

Buffer A tool that resembles a disk sander, but runs at a higher speed and uses a polishing bonnet, to buff the final coat of paint.

Buffing compound An abrasive paste or cake used with a cloth or sheepskin pad to remove fine scratches and to polish lacquer finishes.

Build The amount of paint film deposited, measured in mils.

Bulge A high crown or area of stretched metal.

Bumping The process of smoothing the damaged area following roughing.

Bumping file A tool with a spoon-like shape and serrated surface, used to slap down and shrink high spots.

Bumping hammer A hammer used to roughly pound out a dent.

Bumping spoon A spoon that is often used as a pry bar.

Burnishing The polishing or buffing of a finish by hand or machine using a compound or liquid.

Butt weld A weld made along a line where two pieces of metal are placed edge to edge.

Butyl acetate A solvent for paint, often used in lacquers.

Butyl adhesive A rubber-like compound used to bond fixed glass in place.

Butyl tape Tape that is used with a separate adhesive to bond fixed glass in place.

Caged plate assembly A tapped plate inside a sheet metal box spot welded to the inside of a door or pillar to accept the hinge and striker bolts; is moved in the cage for adjustment.

Calcium A metal component of dryers and pigments.

Calibrate The process of checking equipment to see that it meets test specifications, and that the settings of the equipment are correct.

Camber The inward or outward tilt of a wheel at top from true vertical.

Canister (1) A container of chemicals designed to remove specific vapors and gases from breathing air. (2) A container filled with charcoal used to filter and trap fuel vapors.

Carbonizing flame A welding flame with excess acetylene, which will introduce carbon into the molten metal.

Carbon monoxide (CO) A deadly gas created by the incomplete burning of fuel.

Carnauba wax A hard wax obtained from a species of palm tree and used in some body polishing materials

Carpet The floor covering used in vehicles.

Caster The backward or forward tilt of a kingpin or spindle support arm at top from true vertical.

Casting (1) The process of molding materials using only atmospheric pressure. (2) A part produced by that process.

Catalyst A substance that causes or speeds up a chemical reaction when mixed with another substance but does not change by itself.

Catalytic converter An exhaust component used to reduce harmful exhaust gas emission through a chemical reaction.

Caulking compound A semi-flexible sealer used to fill cracks, seams, joints, and to eliminate water or air leaks and rattles.

C-clamp A C-shaped device with threaded posts used to hold parts together during assembly or welding procedures.

Centering gauge A frame gauge, used in sets of four to locate the horizontal datum planes and centerline on a vehicle.

Centerline strut gauge A device used to detect misalignment of strut towers in relation to the center plane and the datum plane.

Center pillar A box-like column that separates the front and rear doors on a four-door vehicle. Often called a hinge pillar, it holds the front door striker/lock and the rear door hinges.

Center plane An imaginary centerline running lengthwise along the datum plane.

Center punch A tool having a pointed, tapered shaft, used to locate and make starting dents for drilling.

Centimeter A unit of linear measure in the SI metric system.

Ceramic mask An opaque black mask baked directly on the perimeter of the glass to help hide the black polyurethane adhesive and the butyl tape.

Cerium oxide powder An very fine abrasive used to polish out scratches on glass.

Chalking The result of weathering of a paint film characterized by loose pigment particles on the surface of the paint.

Chassis An assembly of mechanisms that make up a major operating system of vehicle, including everything under the body—suspension system, brake system, wheels, and steering system.

Check A small crack.

Check valve A device that only allows the passage of air or fluid in one direction.

Checking A type of failure in which cracks in the paint film begin at the surface and progress downward, usually resulting in a straight V-shaped crack, narrower at the bottom than at the top.

Chemical burn An injury that results when a corrosive chemical strikes the skin or eye.

Chemical shining A spotty discoloration of the topcoat caused by atmospheric conditions, often occurring near an industrial area.

Chipping The breaking away of a small portion of paint film because of its inability to flex under impact or thermal expansion and contraction.

Chlorofluorocarbon A class of refrigerant chemicals that have a detrimental effect on the earth's ozone layer.

Chopping A term that describes the lowering of a vehicle profile.

Chroma The intensity of a color; the degree it differs from the white, gray, or black of the neutral axis of the color tree.

Chronic effect The adverse effect on a human or animal having symptoms that develop slowly over a long period of time or that may reoccur frequently.

Circuit A path for electricity. A circuit must be complete before current will flow.

Circuit breaker A device having a bimetallic strip and a set of contacts that open if excessive current heats the strip to make it bend.

Claimant A person who files a claim.

Clamp saw A saw that is used to cut through spot welds by only cutting through the top layer of material.

Clean To be free of dirt or other material, such as after washing. A bright clear color.

Cleaner A material that is used to clean a substrate.

Clear Transparent. A paint that does not contain pigment or only contains transparent pigment.

Clearance The amount of space between adjacent parts or panels.

Clearcoat A transparent top coating on a painted surface so the base color coat is visible.

Clip A section of a salvaged vehicle. A mechanical fastener used to hold moldings to panel.

Clip removal tool A hand tool used for removing the mechanical fasteners used to attach door handles and trim.

Closed bid One in which the final cost has been determined.

Closed coat abrasive A material in which the abrasive grains are as close together as possible.

Closed structural member A boxed-in section, generally only accessible from the outside, such as a rail or a pillar.

Clouding The formation or the presence of a haze in a liquid or on a film.

Coat, double Two single coats of material, one followed by the other with little or no flash time.

Coat, single A coat of material produced by two passes of a spray gun, one pass overlapping the other in half steps.

Coated abrasive A combination of abrasive grains, backing materials, and bonding agents (sandpaper, for example).

Coating The act of applying paint; the actual film left on a substrate by the paint.

Coatings Covering material used to protect an area.

Cobwebbing The tendency of sprayed paint to form strings or strands rather than droplets as it leaves the spray gun.

Cohesive bonding A process of joining plastics that involves using solvent cements to melt plastic materials together.

Cold knife A device used to cut through adhesive that secures fixed glass or other components.

Cold shrinking The reducing of a surface area of metal by using a shrinking hammer and shrinking dolly.

Cold working The working of metal without the application of heat.

Collision A "wreck" or "crash." Damage caused by an impact on a vehicle body and chassis.

Color The visual appearance of a material, such as red, yellow, blue, or green.

Color holdout The ability of a primer-sealer to allow the finish coat to maintain its high gloss.

Color retention The permanence of a color under a specific set of conditions.

Color sanding The process of wet sanding to remove surface imperfections from the topcoat.

Commission The payment to a technician on the basis of a percentage, usually 40 to 60 percent, of the total labor for a repair job.

Compatibility The ability of two or more materials to be used together.

Composite A plastic formed by combining two or more materials, such as a polymer resin (matrix) and a reinforcement material.

Composite headlight A headlight with a separate bulb, lens, and connector.

Compounding The action of using an abrasive material either by hand or by machine to smooth and bring out the applied topcoat.

Compression resistance spot welding A method using clamping force and resistance heating to form two-sided spot welds.

Computer An electronic device for storing, manipulating, and disseminating information.

Compressor A device used to deliver compressed air for the operation of shop equipment. A device used to circulate refrigerant in an air conditioning system.

Concave An inward curve, like a dent.

Concentration The amount or ratio of any substance in a solution.

Condensation A change in state of a vapor to a liquid on a cold surface, usually moisture. A type of polymerization characterized by the reaction of two or more monomers to form a polymer plus some other product, usually water.

Conductor A material that will allow the flow of electrons.

Connector A device having male and female halves which are fastened together to join a wire or wires.

Consistency The fluidity of a liquid or of a system.

Contaminants The foreign substances on a prepared surface to be painted, or in the paint, that will adversely affect the finish. Any impurities in a material.

Control point The points on a vehicle, including holes, flats, or other identifying areas, used to position panels and rails during the manufacture of the vehicle.

Continuity tester A device that is used to test bulbs, fuses, switches, circuits, and other electrical devices.

Continuous sander A sander that uses an abrasive belt to remove paint and to sand body filler.

Contract An agreement in the form of a written legal document between two or more people.

Conventional frame A vehicle construction type in which the engine and body are bolted to a separate frame.

Conventional point The points on a unibody used as a reference to make a repair.

Conversion coating A chemical treatment used on galvanized steel, uncoated steel, and aluminum to prevent rust.

Convertible top fabric Canvas, synthetic fabric, and vinyl cloth material.

Convex An outward curve, like a bump.

Copper (Cu) A metal. A difficult metal substrate to paint. A metal used in the manufacture of pigments and dryers.

Corrosion A chemical reaction caused by air, moisture, or corrosive materials on a metal surface, referred to as rusting or oxidation.

Corrosion protection The use of any of a variety of methods to protect steel body parts from corrosion and rusting.

Coverage The surface area a given amount of paint will cover. The ability of a topcoat to hide an undercoat.

Core (1) The tubes that form the coolant passages in a radiator or heat exchanger. (2) A rebuildable unit used for exchange when purchasing a new or rebuilt unit.

Corporate Average Fuel Economy (CAFE) The goal of 40 miles per gallon of fuel is to be accomplished by the year 2000 by every automobile manufacturer.

Corporation A business that generally has two or more owners.

Corrosion A breakdown of metal caused by chemical action, such as rust.

Courtesy estimate A written estimate for record-keeping purposes.

Coverage The area that a given amount of paint will cover when applied according to the manufacturer's directions.

Cowl panel A panel forward of the passenger compartment to which the fenders, hood, and dashboard are bolted.

Cowl cut A nose or front end that is cut behind the cowl or firewall.

C-pillar The pillar connecting the roof to the rear quarter panel.

Cracking The splitting of a paint film, usually as straight lines, which penetrate the film thickness, often the result of overbaking.

Crane A portable device that is used to lift or move heavy objects, such as a vehicle engine.

Cratering The formation of holes in the paint film where paint fails to cover due to surface contamination.

Crawling A wet paint film defect that results in the paint film pulling away from, or not wetting, certain areas.

Crazing A paint film failure that results in surface distortion or fine cracking.

Creeper A low platform having four caster rollers used by a technician for mobility when working under a vehicle.

Creeping A condition in which paint seeps under the masking tape.

Cross member A reinforcing piece that connects the side rails of a vehicle frame.

Crown A convex curve or line in a body panel.

Crush zone A section built into the frame or body designed to collapse and absorb some of the energy of a collision.

Cubic centimeter A unit measure of volume in the SI metric system equal to one milliliter.

Cure The process of drying or hardening of a paint film.

Curing The chemical reaction in drying of paints that dry by chemical change.

Current The flow of electrons in an electrical circuit.

Customer requested Repairs that the customer wishes to have made, outside those covered by insurance.

Customize The altering of a vehicle to meet individual tastes or specifications.

Custom-mixed topcoat A special blend of paint that is often used when attempting to match a badly oxidized or faded finish.

Custom paint The refinishing or decorating of a vehicle in a personalized manner.

Cyanoacrylate adhesive system A two-part adhesive that forms an extremely strong bond on hard plastics and similar materials.

Datum line An imaginary line that appears on frame blueprints or charts to help determine the correct height.

Datum plane The horizontal plane or line used to determine correct frame measurements.

Decal Paint films in the form of pictures or letters that can be transferred from a temporary paper backing to another surface.

Dedicated fixture measuring system A bench with fixtures that may be placed in specific points for body measurement.

Deductible The amount of the claim that the vehicle owner must pay, with the remainder paid by the insurance company.

Defroster circuit Narrow bands of a conductive coating that are printed on the interior surface of the rear window glass.

Degradation The gradual or rapid disintegration of a paint film.

Degreasing The cleaning of a substrate, usually metal, by removing greases, oils, and other surface contaminants.

Dehydration The removal of water.

Density The weight of a material per unit of volume, usually grams per cubic centimeter.

Depreciation The loss of value of a vehicle or other property due to age, wear, or damage.

Depression A concave dent.

Detailing The final cleanup and touch-up on a vehicle.

Dew point The temperature at which water vapor condenses from the air.

Diamond A damage condition where one side of the vehicle has been moved to the rear or the front, causing the frame/body to be out of square, or diamond-shaped.

Die A cutting tool used to make external threads or to restore damaged threads.

Diluent A liquid, not a true solvent, used to lower the cost of a paint thinner system.

Dilution ratio The amount of a diluent that can be added to any true solvent when the mixture is used to dissolve a certain weight of polymer.

Dipping Applying paint, primer, or sealer by immersing the part in a container of paint, withdrawing it, and then allowing the excess to drain.

Dinging The process of using a body hammer and a dolly to remove minor dents.

Dinging hammer A hammer used for removing dents.

Direct current (DC) Electrical current that travels in only one direction.

Direct damage Damage that occurs to an area that is in direct contact with the damaging force or impact.

Disassemble To take apart.

Disk grinder An electric or pneumatic tool used with abrasive wheels or disks for heavy-duty grinding, deburring, and for smoothing welds.

Disk sander A rotary power tool used to remove paint and locate low spots in a panel.

Distortion A condition in which a component is bent, twisted, or stretched out of its original shape.

Dirty (1) A color that is not bright and clean, that appears grayish. (2) A condition that requires cleaning.

Dispersion The act of distributing solid particles uniformly throughout a liquid; commonly, dispersion of pigment in a vehicle.

Dispersion coatings Types of paint in which the binder molecules are present as colloidal particles instead of solutions.

Distillation range The boiling temperature range of a liquid.

Doghouse The front clip or front body section, also called the nose section, including everything between the front bumper and the firewall.

Dog tracking The off-center tracking of the rear wheels as related to the front wheels.

DOL An abbreviation for the U.S. Department of Labor.

Dolly block An anvil-like metal hand tool held on one side of a dented panel while the other side is struck with a dinging hammer.

Domestic The classification for any vehicle having 75 percent or more of its parts manufactured in the United States.

Door hinge wrench A specially shaped wrench used to remove or install the retaining bolts on door hinges.

Door lock assembly The assembly that makes up a door lock.

Door lock striker That part of a door lock assembly that is engaged by the latching mechanism when the door is closed.

Door skin The outer door panel.

Double coat The technique of spraying the first pass left to right and spraying the second pass right to left directly over the first pattern.

Drain hole A hole in the bottom of a door, rocker panel, or other component that allows water to drain.

Drier (1) A chemical added to paint to reduce drying or cure time. (2) A system of removing moisture.

Drifting (1) The mixing of two or more colors to achieve the desired final color. (2) A term used for blending.

Drift punch A tool that has a long tapered shaft with a flat tip, used to align holes.

Drill A power tool with interchangeable cutting bits that are used to bore holes.

Drill bit A cutting tool used with a drill.

Drill press A floor- or bench-mounted electric drill used to bore holes.

Drip cap A term used for drip molding.

Drip molding The metal molding that serves as a rain gutter over doors; sometimes called a drip cap.

Drive train The engine, transmission, and drive shaft/axle assembly.

Drop light A portable light attached to an electrical cord used to illuminate a work area.

Dry (1) To change from a liquid to a solid which takes place after a paint is deposited on a surface. (2) To be free of moisture or other liquid.

Dryer (1) A catalyst added to a paint to speed up curing or drying time. (2) A device used for drying material.

Drying The process of changing a coat of paint from a liquid to a solid state due to evaporation of the solvent, a chemical reaction of the binding medium, or a combination of these causes.

Drying time The amount of time required to cure paint or allow solvents to evaporate.

Dry sanding A technique used to sand finishes without use of a liquid.

Dry spray An imperfect coat of paint, usually caused by spraying too far from the surface being painted or on too hot of a surface.

Dual-action sander A sander that combines circular and orbital motion in one device.

Ductility The property that permits metal deformation under tension.

Durability The length of service life; usually applies to a paint used for exterior purposes.

Early-model A term used to describe automobiles that are over 15 years old.

Eccentric An offset cam-like section on a shaft used to convert rotary to reciprocating motion.

E-coat A cathodic electro-deposition coating process that produces a tough epoxy primer.

Edge joint A joint between the edges of two or more parallel or nearly parallel members.

Elastic deformation A condition that occurs when a material is not stretched beyond its elastic limit.

Elasticity A material's ability to be stretched and then return to its original shape.

Elastic limit The point at which a material will not return to its original shape after being stretched.

Elastomer A man-made compound with flexible and elastic properties.

Electrical system The starter, alternator, computers, wires, switches, sensors, fuses, circuit breakers, and lamps used in a vehicle.

Electric convertible top A convertible top that uses an electric motor and actuator assembly for raising and lowering.

Electric-over-hydraulic system A system having an electric motor to drive a hydraulic pump.

Electric tool A tool that operates on electrical power.

Electrocution A condition whereby electricity passes through the human body causing severe injury or death.

Electrode A metal rod used in arc welding that melts to help join the pieces to be welded.

Electronic display A light-emitting diode (LED), digital readout, or other device used to provide vehicle information.

Electronic system A computerized vehicle control system such as the engine control systems or anti-lock brake systems.

Electrostatic spraying The application of paint by high-voltage atomization.

Emulsion A suspension of fine polymer particles in a liquid, usually water.

Enamel A type of paint that dries in two stages: first by evaporation of the solvent and then by oxidation of the binder.

Endless line A custom painting technique in which narrow tape is applied in the desired design.

Energy reserve module An alternate source of power for an air bag system if the battery voltage is lost during a collision.

Entrepreneur One who has his or her own business.

Epoxy A class of resins that may be characterized by their good chemical resistance.

Epoxy fiberglass filler A waterproof fiberglass reinforcement material used for minor rust repair.

Estimating The analyzing of damage and calculating the cost of repairs.

Etching The chemical removal of a layer of base metal to prepare a surface for painting.

Ethylene dichloride A chemical used as a solvent to cement plastic joints.

Ethylene glycol A chemical base that is used for permanent antifreeze.

Evaporation A change in state from a liquid to a gas.

Evaporation rate The speed at which a liquid evaporates.

Expansion tank A small tank used to hold excess fuel or coolant as it expands when heated.

Exposure limit The limit set to minimize occupational exposure to a hazardous substance.

Extended method The replacement of an adhesive material when installing new fixed glass.

Extender pigment An inert, colorless, semi-transparent pigment used in paints to fortify and lower the price of pigment systems.

Extension rod An arm used to connect a lock cylinder to a locking mechanism on a door latch assembly.

Exterior The outside area.

Exterior door handle A device that permits the opening of a door from the outside.

Exterior lock A lock used on front doors and trunk lids or hatches.

Exterior trim piece The moldings and other components applied to the outside of a vehicle.

External-mix gun A paint spray gun that mixes and atomizes the air and paint outside the air cap.

Eye bath A device that is used to flood the eyes with water in case of accidental contact with a chemical or other hazardous material.

Face bar A bare bumper with no hardware attachments.

Face shield A device worn to protect the face and eyes from airborne hazards and chemical splashes.

Factory-mixed top coats Paint that is carefully proportioned at the factory to achieve the desired color and match the original finish.

Factory specifications Specific measurements and other information used during original manufacture of a vehicle.

Fading The loss of color.

Fan The spray pattern of a paint spray gun.

Fanning The use of pressurized air through a paint spray gun to speed up the drying time of a finish.

Fan shroud The plastic or metal enclosure placed around an engine-driven fan to direct air and improve fan action.

Fatigue failure A metal failure resulting from repeated stress that alters the character of the metal so that it cracks.

Feather The tapered edge between a bare metal panel and the painted surface.

Featheredge The tapering of the edge of the damaged area with sandpaper or special solvent.

Featheredge splitting Stretch marks or cracks along the featheredge that occur during drying or shortly after the topcoat has been applied over a primed surface.

Feathering The act of using sandpaper to taper the paint surface around a damaged area, from the base metal to topcoat.

Fiberglass A material composed of fine-spun filaments of glass used as insulation, and for reinforcement of a resin binder when repairing vehicle bodies.

Fiberglass cloth A heavy woven reinforcement material that provides the greatest physical strength of all the fibrous mats.

Fibrous composites A material that is composed of fiber reinforcements in a resin base.

Fibrous mat A reinforcing material consisting of nondirectional strands of chopped glass held together by a resinous binder.

Fibrous pad A pad of resin fiber or fiberglass that is used on the inside of the hood and other areas to deaden sound and provide thermal insulation.

File A tool with hardened ridges or teeth cut across its surface used for removing and smoothing metal.

Filler A material that is used to fill a damaged area.

Filler metal The metal added when making a weld.

Filler panel A panel that is found between the bumper and the body.

Filler strip A strip included in windshield installation kits to be used in the antenna lead area.

Film A very thin continuous sheet of material, such as paint that forms a film on the surface to which it is applied.

Finish (1) A protective coat of paint. (2) To apply a paint or paint system.

Finish coat The final coat of finish material applied to a vehicle.

Finishing hammer A hand tool used to hammer metal.

Finish sanding (1) The last stage in hand sanding the old finish. (2) Sanding the primer-surfacer using #400 grit or finer sandpaper.

Fish-eye A paint surface depression in wet paint film usually caused by silicone contamination of the paint.

Fish-eye eliminator An additive that makes paint less likely to show fish-eye.

Fixed glass Glass, such as a rear window, that is not designed to move.

Fixed pricing Standard pricing for performing a service or repair that does not change.

Fixture An accessory for a dedicated measuring system designed to attach to a bench to fit reference points.

Flaking A paint failure noted by large pieces of paint separating from the substrate.

Flames A flame-like design produced by the use of stencils and paint.

Flash The first stage of drying where some of the solvents evaporate, which dulls the surface from an exceedingly high gloss to a normal gloss.

Flashback A condition in which the oxygen-acetylene mixture burns back into the body of the welding torch.

Flasher unit An electrical device used to flash the turn signal and hazard lights.

Flash point The temperature at which the vapor of a liquid will ignite when a spark is struck.

Flash time The time between coats or paint application and/or baking.

Flat The lack of a gloss or shine.

Flat boy A term often used for a speed file.

Flat chisel A tool designed for shearing steel, including removing bolts or rivets.

Fat rate A predetermined time allowed for a particular repair and the money charge to perform that repair based on a standard per hour shop fee.

Flattener An additive that reduces the gloss of a finishing material.

Flex-additive A material added to a topcoat to make it flexible.

Floor jack Portable equipment used to lift a vehicle.

Floor pan The main underbody assembly of a vehicle that forms the floor of the passenger compartment.

Flow (1) The leveling characteristics of a wet paint film. (2) The ability of a liquid to run evenly from a surface and to leave a smooth film behind.

Fluid adjustment screw A spray gun control that is used to regulate the amount of material passing through the fluid tip as the trigger is depressed.

Fluid control valve A manual spray gun adjustment used to determine the amount of paint coming from the gun.

Fluid needle A spray gun valve component that shuts off the flow of material.

Fluid tip A spray gun nozzle that meters the paint and directs it into the air stream.

Fog coat A paint coat following a wet coat in which mottling or streaking occurs.

Foreign A general classification for a vehicle that has less than 75 percent of its parts manufactured in the United States.

Frame The heavy metal structure that supports the auto body and other external components.

Frame alignment The act of straightening a frame to the original specifications.

Frame-and-panel straightener A portable or stationary hydraulically powered device used to repair damaged sheet metal and correct frame damage.

Framed door A design in which the door frame surrounds and supports the glass.

Frame gauge A gauge that may be hung from the car frame to check alignment.

Frame straightener A pneumatic- or hydraulic-powered device used to align and straighten a distorted frame or body.

Frame system A heavy frame assembly upon which the various attachments are mounted.

Free air capacity The actual amount of free air that is available at the compressor's working pressure.

Frisket paper An adhesive-backed paper used as stencil material.

Front body hinge pillar The pillar to which the front door hinges are attached.

Front-wheel drive A vehicle that has its drive wheels located on the front axle.

Frosting (1) The formation of a surface haze. (2) The defects in a drying paint film. (3) The freezing of surface moisture on a line or component.

Full cut-out method The replacement of an adhesive material when installing new fixed glass.

Full body section A section repair to both rocker panels, windshield pillars, and floor pan that is required to join the undamaged front half of one vehicle to an undamaged rear half of another vehicle.

Full frame The strong, thick steel structure that extends from the front to the rear of vehicle.

Full wet coat A heavy application of finish used to thoroughly cover the substrate.

Fuse block A panel-like holder for fuses and circuit breakers to the vehicle's electrical circuits.

Fuse An electrical protective device with a soft metal element that will melt and open an electrical circuit if more than the rated amount of current flows through it.

Fusibility A measure of a material's ability to join another while in a liquid state.

Fusible link A specially designed wire joint that melts and opens the circuit if excessive current flows.

Fusion weld A joining operation that involves the melting of two pieces of metal together.

Galvanized metal Metal that is coated with zinc.

Gap The distance between two points.

Garnish molding A decorative or finish molding around the inside of glass.

Gasket (1) A rubber strip used to secure fixed glass on early-model vehicles. (2) A cork, rubber, or combination of the two used as a seal between two mating surfaces.

Gas metal arc cutting (GMAC) An arc cutting process used to sever metals by melting them with the heat of an arc between a continuous metal electrode and the work.

Gas metal arc welding (GMAW) Also called MIG welding, an arc welding process that produces coalescence of metals by heating them with an arc between a continuous filler metal electrode and the work.

Gauge (1) A measure of thickness of sheet metal. (2) A device used to indicate a system condition, such as pressure or temperature.

General-purpose file A flat, round, or half-round shape tool used to remove burrs and sharp edges from metal parts.

General purpose hammer A hammer used for striking tools or tasks other than shaping sheet metal.

General purpose tool A tool that is common to any shop that performs automotive service or repair.

Glass installer The technician responsible for windshield and door glass replacement.

Glass run channel A term used for window channel.

Glass spacer The rubber pieces used to position or align glass.

Glazing putty A paste-like material used to fill small surface pits or flaws.

Gloss The ability of a surface to reflect light as measured by determining the percentage of light reflected from a surface at certain angles.

Goggle Glasses having colored lenses or clear safety glass that protects the eyes from harmful radiation during welding and cutting operations.

Grain pattern The surface appearance and color variation of vinyl fabric.

Grater file A file used to shape body filler before it has completely hardened.

Gravity-feed gun A spray gun into which paint is fed by gravity.

Greenhouse The passenger portion of a vehicle body.

Grille The decorative panel in front of the radiator.

Grit A measure of the size of the particles on sandpaper or disc.

Grommet A donut-shaped rubber device used to surround wires or hoses for protection where they pass through holes in sheet metal.

Ground-return system In metal-framed vehicles, the frame is part of the electrical circuit, so only one wire is needed to complete the circuit. Composite or plastic bodies require two wires.

Guide coat A reference coat of a different color often applied to a primer-surfacer to be sanded off to visually determine if the panel is straight.

Hacksaw A hand saw used to cut metal.

Hand rubbing compound A rubbing compound designed for manual use only.

Hardener A curing agent used in certain plastics and epoxies.

Hard-faced hammer A hammer used to strike tools or to bend or straighten metal parts.

Hard hat A metal or plastic headgear worn to help protect the head from abrasions, hot sparks, and chemical sprays.

Hardness The quality of a dry paint film that gives film resistance to surface damage or deformation.

Hardtop An automobile body style that does not have a roof-supporting center pillar.

Hardtop door A door design without a frame around the glass that rests against the top and sides of the door opening and the weather stripping.

Hardware (1) Computer, printers, hard drives, CD-ROM drives, and other computer equipment. (2) Hinges, hangers, and fasteners used in vehicle construction.

Harness The electrical wires and cables that are tied together as a unit.

Hazardous material Any material that can cause serious physical harm or pose a risk to the environment.

Hazardous substance Any hazardous material that poses a threat to waterways and the environment.

Hazardous waste Any material that can endanger human health if handled or disposed of improperly.

Haze The development of a cloud in a film or in a clear liquid.

Header bar The framework or inner construction that joins the upper sections of the windshield or the back glass and pillars to form the upper portion of the windshield or the back glass opening, and that reinforces the top panel.

Head liner A cloth or plastic material covering the roof area inside the passenger compartment.

Heat gun An electric hand-held tool that blows heated air to soften plastic parts or for speed drying.

Heat shrink The heating with a torch, then using a hammer and dolly to flatten a panel, and then quenching it with water to shrink the metal.

Hem flange A flange at the bottom of a door panel.

Hemming tool A pneumatic tool used to create a hem seam.

Hem seam A door bottom seam formed by bending the outer panel hem flange over the inner panel flange with a hammer and a dolly.

Hiding The degree to which a paint obscures the surface to which it is applied.

High-solids systems A system that uses a high-volume low-pressure spray gun.

High-strength low alloy A type of steel used in unibody design manufacture.

High-strength steel A low-alloy steel that is stronger than hot- or cold-rolled steel; used in the manufacture of structure parts.

High-volume low-pressure gun A spray gun that atomizes paint into low-speed particles.

Hinge pillar The framework or inner construction to which a door hinge is fastened.

Hold out The ability of a surface to keep the topcoat from sinking in.

Hooding To apply a cover or hood to the headlight area that is constructed of epoxy resin and fiberglass reinforcement.

Hood panel A large metal panel that fills the space between the two front fenders and closes off the engine compartment.

Hot gas welding The use of air or inert gas that is heated by a torch to melt and fuse thermoplastics and plastic filler rods together.

Hot glue gun An electrically heated device that is used to melt and apply adhesive to plastics and other materials.

Hot knife An electrically heated tool that is used for cutting the polyurethane adhesive that is used to secure windshields.

Hot-melt A polymer adhesive that is applied in its molten state.

Hotspot An unprotected area that may be subject to corrosion.

Hot spray A technique of applying hot paint.

HSLA An abbreviation for high-strength low-alloy steel.

HSS An abbreviation for high-strength steel.

Hue A visual characteristic by which one color will differ from another, such as red, blue, and green.

Humidity The amount of water vapor in the air.

Hydraulic The use of a fluid under pressure to do work.

Hydraulic-electric convertible top A system that uses hydraulic cylinders and an electric motor to raise and lower the top.

Hydraulic press A press having a hydraulic jack, or cylinder, that is used for pressing, straightening, assembling, or disassembling components.

Hydraulic tool A tool having a pump system that forces hydraulic fluid into a cylinder to push or pull a ram.

Hydrometer An instrument used to measure the specific gravity of fluids, such as battery electrolyte.

Hydrocarbon A compound that contains carbon and hydrogen.

I-CAR An acronym for Inter-Industry Conference on Auto Collision Repair.

Impact chisel An electrically or pneumatically driven hand tool that creates a hammering and reciprocating action on a chisel bit used to cut metal.

Impact tool An electric or pneumatic driven hand tool used to tighten or loosen bolts and nuts.

Impact wrench An electrically or pneumatically driven hand tool that is used to tighten or loosen nuts and bolts.

Impurities Foreign material, such as paint, rust, or other contaminants, that can substantially weaken a weld joint.

Included angle An angle that places the turning point of the wheel at the center of the tire-to-road contact area.

Independent garage A term often used for an independent shop.

Independent shop A repair shop that may be a sole proprietorship or a partnership.

Independent front suspension A conventional front suspension system in which each front wheel moves independently of the other.

Independent rear suspension A rear suspension system that has no cross axle shaft and each wheel acts independently.

Induction heating The generation of heat in a substrate by the application of an electromagnetic field.

Indicator lamp (1) A bulb used to warn of problems with oil pressure, engine temperature, fuel level, or alternator output. (2) A bulb used for turn signals or hazard signals.

Indirect damage Any damage that occurs away from the point of impact.

Infrared baking The drying of a paint film using heat developed by an infrared source.

Infrared dryer An electrical heating element that emits radiant energy for the drying or curing of automotive finishes.

Infrared light A portion of the spectrum that accounts for most of the heating effects of the sun's rays.

Inhibitor An additive for paint that slows gelling, skinning, or yellowing processes.

Inner panel An automotive body component that adds strength and rigidity to the outer panels.

Install To attach or insert a part, component, or assembly to a vehicle.

Instructor A professional who teaches others automotive mechanics, automotive body repair and refinishing, or any other trade.

Insulation A material that is commonly applied to muffle excessive noise, reduce vibration, and prevent unwanted heat transfer.

Insulator Any material that opposes the flow of electricity.

Insurance adjuster One who reviews estimates to determine which best reflects how the vehicle should be repaired.

Insurance adjustment An agreement between the vehicle owner, insurance company, and the body shop regarding what repairs will be made and who will pay for them.

Integrally welded A term that describes two or more parts that are welded together to form one integral unit.

Interchangeability The ability of new or used replacement parts to fit as well as the original manufactured part.

Interior door trim assembly The coverings and hardware in the cabin surface of a door.

Interior lock A mechanical lever, knob, or button attached via a rod to the lock portion of the latch assembly.

Interior trim All of the upholstery and moldings on the inside of the vehicle.

Internal-mix gun A spray gun that mixes air and material inside the cap.

Internal rust-out Rust damage caused by oxidization from inside to the outside.

Iron A basic element.

Isocyanate resin The principal ingredient in urethane hardeners.

Isopropyl alcohol A solvent that will dissolve grease, oil, and wax, but will not harm paint finishes or plastic surfaces.

Jack A device used for heavy lifting, such as a vehicle.

Jack stand A safety device used to support the vehicle when working under it.

Jig A mechanical device used for positioning and holding work.

Joint The area at which two or more pieces are connected.

Jounce The compression of a spring caused by an upward movement of the wheel and/or a downward movement of the frame.

Jumper wire An electrical test component used to connect or bypass a component for testing.

Jump start The act of connecting a vehicle with a dead battery to a good battery, so that enough current will flow to start the engine.

Jump suit A garment that resists paint absorption, provides full-body protection, and can be worn over other clothing.

Kerf The space left after metal has been removed by cutting with a saw or torch.

Keyless lock system A lock system that operates the lock through use of a numeric keypad on which a code is entered, or by a signal from a small transmitter.

Kick-out The precipitation of the dissolved binder from a solution as a result of solvent incompatibility.

Kick pad The panel that fits between the cowl and the front door opening.

Kick over A term used by some to indicate that a plastic filler has hardened.

Kick panel The panel that fits between the cowl and the front door opening.

Kilogram A unit of measure in the SI metric system.

Kilometer A unit of measure in the SI metric system.

Kink A bend of more than 90° in a distance of less than 0.118 inches (3 mm).

Kinking A method of cold shrinking by using a pick hammer and a dolly to create a series of pleats in the bulged area.

Lace painting A stencil painting technique in which paint is sprayed over fabric lace designs that have been stretched across the surface to be decorated.

Lacquer A type of paint that dries by solvent evaporation and can be rubbed to improve appearance.

Ladder frame A frame design in which the rails are nearly straight with the cross members to stiffen the structure.

Laminar composite Several layers of reinforcing materials that are bonded together with a resin matrix.

Laminated glass Any glass having a plastic film sandwiched between layers for safety.

Lamination A process in which layers of materials are bonded together.

Lap weld A weld that is made along the edge of an overlapping piece.

Laser system A type of measuring system that uses laser optics.

Latch A mechanism that grasps and holds doors, hoods, and trunk lids closed.

Latch assembly A manual- or power-operated handle mechanism for the trunk or hatch.

Late-model A vehicle manufactured within the past 15 years.

Leader hose A short length of air hose with quick coupler connections used to connect pneumatic tools to the shop air supply.

Leading The act of applying a lead-based solder body filler.

Liability A legal responsibility for business decisions and actions.

Liability insurance Insurance that covers the policy-holder on liability damages to the personal property of others.

Lift channel A channel in which window glass is supported as it is raised and lowered.

Lifting The attack of an undercoat by the solvent in a top coat, resulting in distortion or wrinkling of the undercoat.

Lift A hydraulic mechanism used to raise a vehicle off the floor.

Light bulb An electrical device with internal elements that glow when electrical energy is applied.

Lightness The whiteness of paint measured by the amount of light reflected off its surface.

Liquid sandpaper A chemical that cleans and etches the paint surface.

Liquid vinyl (1) A paint that consists of vinyl in an organic solvent. (2) Material sometimes used to repair holes or tears in vinyl upholstery or similar applications.

Liquid vinyl patching compound A thick material that may be used to repair severely damaged areas of vinyl.

Listing A small pocket in the headlining that holds support rods.

Liter A unit of measure SI metric system.

Load Any device or component that uses electrical energy.

Load tester An instrument used to determine the condition of a battery.

Lock A mechanism that prevents a latch assembly from operating.

Lock cylinder The mechanism operated with a key in a mechanical lock system.

Locking cord A term often used for locking strip.

Locking strip The strip that fits into the gasket groove to secure the glass in the gasket.

Lock-out tool A device used to open a door if a door latch is inoperative or the keys are lost or locked inside.

Lock pillar The vertical door post containing the lock striker.

Longitudinal A term generally used to identify an engine that is positioned so the crankshaft is perpendicular to the vehicle's axles.

Lord Fusor The trade name for a body panel repair adhesive recommended for bonding panels to space frames.

Low crown A damage area with a slightly convex curve.

Lower glass support A support placed at the bottom edge of the window opening.

Low spot A small concave dent.

Machine guard A safety device used to prevent one from coming into contact with the moving parts of a machine.

Machine rubbing compound A compound with very fine abrasive particles, suitable for machine application.

MacPherson strut A type of independent suspension that includes a coil spring and a shock absorber.

Maintenance-free battery A battery designed to operate its full service life without requiring additional electrolytes.

Major damage Any damage that includes severely bent body panels and damaged frame or underbody components.

Malleability The property that permits metal formation and deformation under compression.

Manager/Supervisor One who has control of shop operations and the hiring, training, promotion, and firing of personnel.

Manual convertible top A top system that is raised and lowered by hand.

Manually operated seat A seat system that is manually adjusted back and forth on a track.

Manual welding A welding process in which the procedure is performed and controlled by hand.

Markup The amount of profit that is added to the cost to determine selling price.

Marred A part or component that has been damaged or scratched.

Mash A vehicle body damage in which the length of any section or frame member is less than factory specifications.

Masking Paper or plastic used to protect surfaces and parts from paint overspray.

Masking paper A special paper that will not permit paint to bleed through.

Masking tape An adhesive-coated paper-back tape used to protect parts from spray paint.

Mass tone The color of paint as it appears in the can or on the painted panel.

Material safety data sheet (MSDS) Data that are available from all product manufacturers detailing chemical composition and precautionary information for all products that can present a health or safety hazard.

Matting Glass-fiber materials loosely held together and used with a resin to make repairs.

Measurement system A system that allows one to check for frame or body alignment or misalignment.

Mechanical fastener A device, such as screws, nuts, bolts, rivets, and spring clips, for the adjustment and replacement of assemblies or components.

Mechanical joining The technique for holding components together through the use of fasteners, folded metal joints, or other means.

Mechanical measuring system A system having a precision beam and tram-like adjustable pointers to verify dimensions.

Mechanism The working parts of an assembly.

Metal conditioner A chemical cleaner used to remove rust and corrosion from bare metal that helps prevent further rusting.

Metal inert gas (MIG) welding A welding technique that uses an inert gas to shield the arc and filler electrode from atmospheric oxygen.

Metal insert A component used to help strengthen and secure the repair when sectioning rocker panels.

Metallic Paint finishes that include metal flakes in addition to pigment.

Metallic paint finish Paint that contains metallic flakes in addition to pigment.

Metallurgy The study of metals and the technology of metals.

Metal snips A hand-held scissor-like tool used to cut thin metal.

Metamerism Two or more colors that match when viewed under one light source, but do not match when viewed under a second light source.

Meter (1) An instrument used to make measurements. (2) A unit of measure in the SI metric system.

Mica A color pigment or particle found in pearl paints.

MIG An acronym for metal arc welding.

MIG spot welding A technique often used to tack panels in place before welding with continuous MIG welds or compression resistance spot welding.

Mil A measure of paint film thickness equal to one one-thousandth of an inch.

Mildew A fungus growth that appears in warm, humid areas.

Mill-and-drill pad An attachment point used when sectioning a plastic body panel.

Millimeter A unit of measurement in the SI metric system.

Mineral spirits A petroleum-based product having about the same evaporation rate as gum turpentine; sometimes used for wet sanding and to clean spray guns.

Minor damage Any repair that requires relatively little time and skill to complete.

Mirror bracket adhesive A strong bonding material used to mount a rear-view mirror on the inside of a windshield.

Misaligned Uneven spacing, as between body panels.

Miscible Capable of being mixed.

Mist Liquid droplets suspended in the air due to condensation from the vapor to liquid state, or by breaking up a liquid into a dispersed state by atomizing.

Mist coat A light spray coat of high-volume solvent for blending and/or gloss enhancement.

Model year The production period for new model vehicles or engines.

Mold core method A procedure used to repair curved or irregularly shaped sections.

Molecule The smallest possible unit of any substance that retains characteristics of that substance.

Monocoque A unibody vehicle construction type in which the sheet metal of the body provides most of the structural strength of the vehicle.

Motorized seat belt A seat belt system designed to automatically apply the shoulder belts to the front-seat passengers.

Mottling A paint film defect that appears as blotches or surface imperfections.

Molding clip A mechanical fastener used to secure trim.

Movable glass A window designed to be moved up and down or side-to-side for opening and closing.

Movable section A component held in position by a mechanical fastener.

Mud A slang term used for ready-to-use plastic filler.

Multimeter An electrical instrument that is used to measure resistance, voltage, and amperage.

Multiple-pull system A system that pulls in two or more directions to correct damage.

Music wire Steel wire used for cutting through adhesives.

Negative caster A condition occurring when the top of the steering knuckle is tilted toward the front of the vehicle.

Negligent To be careless or irresponsible.

Net The amount of money left after paying all overhead expenses; known as profit.

Neutral flame The flame of an oxyacetylene torch that has been adjusted to eliminate all of the inner cone acetylene feather.

Neutralizer Any material used to chemically remove any trace of paint remover before finishing begins.

Nibbler A power hand tool used to cut small bites out of sheet metal.

Nitrile glove Gloves used for protection when working with paints, solvents, catalysts, and fillers.

No-fault insurance A type of insurance that covers only the insured's vehicle and/or personal injury, regardless of who caused the accident.

Noise intensity The loudness of a noise.

Non-bleeder gun A paint spray gun having a valve that shuts off the airflow when the trigger is released.

Noncompetitive estimate A detailed and accurate estimate usually done for minor damage where no claims are to be filed with an insurance company.

Nonferrous metal A metal that contains no iron, such as aluminum, brass, bronze, copper, lead, nickel, and titanium.

Nose The front body section ahead of the doors, including bumper, fenders, hood, grille, radiator, and radiator support.

OEM An abbreviation for original equipment manufacturer.

Office staff Those who perform office duties such as billing, receiving payments, making deposits, ordering parts, and paying bills.

Off-the-dolly dinging A technique of holding the dolly away from the raised areas being hit by the hammer.

Ohm A unit of measure for resistance.

Ohmmeter An electrical device used to measure resistance.

Oil A viscous liquid lubrication product derived from various natural sources, such as vegetable oil.

One-wire system The electrical wiring system for a vehicle that uses the chassis as an electrical path to ground, eliminating the need for a second wire.

On-the-dolly dinging A technique of holding the dolly directly under the area where the hammer is used.

On-the-job training A method whereby the beginning technician learns the trade from experienced technicians while taking part in hands-on repairs.

Opaque Not transparent, or impervious to light.

Open bid The scenario in an estimate whereby a part may be suspect of needed repair or replacement but cannot be determined until the repairs are under way.

Open circuit An incomplete circuit due to a break or other interruption that stops the flow of current.

Open-coat abrasive An abrasive in which the abrasive grains are widely separated.

Open structural member A flat panel accessible from either side, such as a floor panel.

Orange peel An irregularity in the surface of a paint film that appears as an uneven or dimpled surface but feels smooth to the touch.

Orbital sander A hand-held power sanding tool that operates in an elliptical or oval pattern.

Orifice A small calibrated opening.

Original finish The paint that is applied by the vehicle manufacturer.

Outer-belt weatherstrip The material that is located between the door panel and the window glass to prevent dirt, air, and moisture from entering.

Outer panel The sheet metal section that is attached to an inner panel to form the exterior of a vehicle.

Outside surface rust Rust that starts on the outside of a panel.

Oven Equipment that is used to bake on a finish.

Overage The added charges for any damage that may be discovered after the original estimate.

Overall repainting Refinish repair that includes completely repainting the vehicle.

Overhaul A procedure in which an assembly is removed, cleaned, and/or inspected, and damaged parts are replaced, rebuilt, and reinstalled.

Overhead dome light Lights that provide illumination in the interior of a vehicle.

Overlap (1) The spray that covers the previous spray stroke. (2) In estimating, when two operations share common steps or procedures, thereby the same flat-rate charges.

Overlay A thin layer of decorative plastic material often with a design or pattern applied to various parts of the vehicle.

Pitting The appearance of holes or pits in a paint film while it is wet.

Plasma A gas that is heated to a partially ionized condition enabling it to conduct an electric current.

Plasma arc cutting A cutting process in which metal is severed by melting a localized area with an arc and removing the molten material with a high velocity jet of hot, ionized gas.

Plastic A manufactured lightweight material that is now being used in automobile construction.

Plastic alloy A material that is formed when two or more different polymers are mixed together.

Plastic deformation The use of compressive or tensile force to change the shape of metal.

Plastic filler A compound of resin and fiberglass used to fill dents and level surfaces.

Plasticity The property that allows metal to be shaped.

Plasticizer A material that is added to paint to make film more flexible.

Plastic-resin mixture A material used to fill chips and pits on a windshield.

Platform frame A frame that consists of a floor pan and a central tunnel.

Pliers A hand tool designed for gripping.

Pliogrip A body repair adhesive used for bonding panels to space frames.

Plug weld The adding of metal to a hole to fuse all metal.

Pneumatic flange tool An air operated hand tool used to form an offset crimp along the joint edge of a panel.

Pneumatic tool A tool that is powered by compressed air.

Polisher A term used for buffer.

Polishing compound A fine abrasive paste used for smoothing and polishing a finish.

Polishing cream An extremely fine abrasive material used for manual removal of swirl marks left after machine compounding.

Polyblend A plastic that has been modified by the addition of an elastomer.

Appendices
Glossary

Polyester resin A thermosetting plastic used as a finish and matrix binder with reinforcing materials.

Polyethylene A thermoplastic used for interior applications.

Polypropylene A thermoplastic material used for interior and some underhood applications.

Polyurethane A chemical compound that is used in the production of resins for enamels.

Polyurethane adhesive A plastic compound used with butyl tape to bond fixed glass in place.

Polyurethane enamel A refinishing material that provides a hard, tile-like finish.

Polyurethane foam A plastic material used to fill pillars and other cavities, adding strength, rigidity, and sound insulation.

Polyurethane primer A material that may be brushed onto the areas where adhesive will be applied to hold glass in place.

Polyvinyl chloride A thermoplastic material used in pipes, fabrics, and other upholstery materials.

Pool An area in molten metal that is created by the heat of the welding process.

Poor drying A condition in which a finish stays soft and does not dry or cure as quickly as the painter may like.

Pop rivet gun A hand-held tool designed to place and secure rivets into a blind hole.

Porosity Voids or gas pockets in any material.

Portable alignment system An alignment system used for correcting frame and body damage.

Portable grinder A hand-held tool used for grinding.

Positive caster The condition that results when the top of the steering knuckle is tilted toward the rear of the car.

Positive post The positive terminal of a battery.

Pot life The time a painter has to apply a plastic or paint finish to which a catalyst or hardener has been added before it will harden.

Pot system A rail alignment system that uses portable hydraulic units anchored to attachment holes in the floor.

Power ram A hydraulic body jack used to correct severe damage.

Power ratchet An electrical or pneumatic powered tool used to remove and replace nuts, bolts, and other fasteners.

Power seat A seat that may be adjusted vertically and horizontally with the use of electric motors.

Power source A source of electrical energy, such as the battery.

Power tool A tool that operates off electrical, hydraulic, or pneumatic power.

Power washer A cleaning machine that uses a high-pressure spray of water to dislodge debris.

PPM An abbreviation for parts per million.

Press fit A joining technique in which one part is forced into the other.

Pressure A force measured in pounds per square inch (psi) or kiloPascals (kPa), such as the air delivered to a paint spray gun.

Pressure drop A loss of air pressure between the source and the point of use.

Pressure-feed gun A paint spray gun in which air pressure or a high-pressure pump force the paint to the gun.

Pressure pot A paint spray system in which paint is fed to the spay gun by air pressure.

Pressurize To apply a pressure that is greater than atmospheric pressure.

Primary damage The damage that occurs at the point of impact on a vehicle.

Prime coat The first coat to improve adhesion and provide corrosion protection.

Primer A type of paint that is applied to a surface to increase its compatibility with the topcoat and to improve adhesion or corrosion resistance.

Primer-sealer An undercoat that improves the adhesion of a topcoat and seals old painted surfaces that have been sanded.

Primer-surfacer A high-solids sandable primer that fills small voids and imperfections.

Priming A process to smooth the surface and help the paint topcoat to bond.

Proprietorship A business, such as a body shop, that is owned by one person.

Puller (1) A tool used to pull out dents. (2) A tool used to remove hubs and pulleys.

Pulling Applying a force.

Pull rod A tool that allows repairs to be performed from the outside of a damaged panel.

Pull tab A metal tab welded to a damaged panel that allows the use of a slide hammer.

Putty A material that is used to fill flaws.

PVC (1) An abbreviation for polyvinyl chloride, a type of plastic. (2) An abbreviation for pigment volume content, the percentage of pigment in solid material of a paint.

Quarter panel The side panel extending from the rear door to the end of a vehicle.

Quench To cool quickly.

Rack-and-pinion steering A steering gear in which a pinion gear on the end of the steering shaft meshes with a rack gear on the steering linkage.

Rail The major member that forms the box-like support in unibody construction.

Rail system A specially designed steel member set into the shop floor providing an anchorage for pushing and pulling equipment.

Reaction injection molding A process involving injecting reactive polyurethane or a similar resin onto a mold.

Rear clip The rear portion of the car including part of the roof.

Rear compartment lid The trunk lid or panel and reinforcement that covers the luggage compartment.

Rear-engine An engine that is positioned directly above or slightly in front of the rear axle.

Rechrome To replace a part, such as a bumpers, with chrome.

Reciprocating sander A hand-held power sander with a sanding surface that moves in small circles while moving in a straight line.

Reduce To lower the viscosity of a paint by the addition of solvent or thinner.

Reducer A solvent combination used to thin enamel.

Reference point The point on a vehicle, including holes, flats, or other identifying areas, used to position parts during repairs.

Refinish To repaint by removing or sealing an old finish and applying a new topcoat.

Reflow A process by which lacquers are melted to produce better flow characteristics.

Regulator (1) A mechanism used to raise or lower glass in a vehicle door. (2) A device used to control pressure of liquids or gases.

Reinforced reaction injection molding A process that involves injecting reactive polyurethane, polyurea, or dicyclopentadiene resin into a mold that contains a preformed glass mat.

Reinforcement piece A sheet metal welded in place along a joint to strengthen the joint.

Reinforcements Structural braces used to strengthen panels.

Relative density The mass of a given volume compared to the same volume of water at the same temperature, referred to as specific gravity.

Relief valve A safety valve designed to open at a specified pressure.

Remove and reinstall A term for removing an item to gain access to a part, then reinstalling the item.

Remove and repair A term for removing and reinstalling a part.

Replacement panel A body panel used to replace a damaged panel.

Resin A term used for polyester or epoxy resin.

Resistance An opposition to the flow of electricity.

Resistance weld A weld made by passing an electric current through metal between the electrodes of a welding gun.

Resource Conservation and Recovery Act A law passed to enable the Environmental Protection Agency to control, regulate, and manage hazardous waste generators.

Respirator A mask worn over the mouth and nose to filter out particles and fumes from the air being breathed.

Retaining clip A spring-type device used to secure one component to another.

Retaining strip A strip sewn into the head liner and attached to T-slots on the inner roof panel to support the head liner.

Retarder A slow evaporating additive used to slow drying.

Reveal file A curved file used to shape tight curves or rounded panels.

Reveal molding An exterior trim piece used to accent a glass opening.

Reverse masking A spot repair technique used to help blend the paint and make the repair less noticeable.

Reverse polarity The connecting of a MIG welding gun to the positive terminal of the DC welder providing greater penetration.

Right-to-know law One's right to essential information and stipulations for safely working with hazardous materials.

Rocker panel A narrow panel attached below the car door that fits at the bottom of the door opening.

Roof rail The framework or inner construction that reinforces and supports the sides of the roof panel.

Roughing out The preliminary work of bringing the damaged sheet metal back to the approximate original contour.

Rubber stop An adjustable hood bumper used to make minor alignment adjustments.

Rubbing compound An abrasive that smooths and polishes paint film.

Runs and sags A heavy application of sprayed material that fails to adhere uniformly to the surface.

Rust Corrosion that forms on iron or steel when exposed to air and moisture.

Rust inhibitor A chemical applied to steel to retard rusting or oxidation.

Rust-out A condition that occurs when rust is allowed to erode through a metal panel.

SAE An abbreviation for the Society of Automotive Engineers.

Safety glasses Protective eyewear.

Safety shoes Protective footwear.

Safety stand A metal support that is placed under a raised vehicle.

Sag Frame damage in which one or both side rails are bent and sag at the cowl.

Sagging Excessive paint flow on a vertical surface resulting in drips and other imperfections.

Salary A fixed dollar amount that is paid a worker per day, week, month, or year.

Sales personnel A manufacturer or equipment suppliers representative who sells various products or services.

Salvage The value of a wrecked vehicle that has been declared beyond repair.

Sandblasting A method of cleaning metal using an abrasive, such as sand, under air pressure.

Sander A hand-held power tool used to speed the rate of sanding or polishing surfaces.

Sanding The use of an abrasive coated paper or plastic backing to level and smooth a body surface being repaired.

Sanding block A hard flexible block that provides a smooth backing for hand sanding operations.

Sand scratches Marks that are made in metal or an old finish by an abrasive.

Sandscratch swelling A swelling of sandscratches in a surface caused by solvents in the topcoat.

Saturate To fill an absorbent material with a liquid.

Scanner An electronic device for reading and storing data or computer information.

Scraper A hand-held tool used to scrape away paint or other surface material.

Scratch awl A pointed tool used for marking and piercing sheet metal.

Screwdriver A hand-held, pneumatically- or electrically-powered tool used to tighten or loosen screw-type fasteners.

Scuff To roughen a surface by rubbing lightly with sandpaper to provide a suitable surface for painting.

Sealed beam headlight A light in which the filament, reflector, and lens are fused into a single hermetic unit.

Sealer A coat between the topcoat and the primer or old finish to give better adhesion.

Sealing strip A strip inside the door that prevents dust from entering the drain holes while allowing water to drain.

Seat belt A restraint that holds occupants in their seats.

Secondary damage The indirect damages that may occur due to misplaced energy that causes stresses at areas other than the primary impact zone.

Sectioning The act of replacing partial areas of a vehicle.

Seeding The development of small insoluble particles in a container of paint which results in a rough or gritty film.

Self-centering gauge A device used to show misalignment.

Self-contained respirator A compressed air cylinder equipped respirator that provides protection.

Self-etching primer A primer that contains an agent that improves adhesion.

Semi-gloss A gloss level between high gloss and low gloss.

Series circuit An electrical circuit with one or more loads wired so the current has only one path to follow.

Service manual A book published by the vehicle manufacturer that lists specifications and service procedures.

Setting time The time it takes solvents to evaporate or resins to cure or become firm.

Settling The separation caused by gravity of one or more components from a paint that results in a layer of material at the bottom of a container.

Shaded glass Glass having a dark color band across its top portion.

Shading A custom painting technique accomplished by holding a mask or card in place and overspraying the surrounding area.

Sheen The gloss or flatness of a film when viewed at a angle.

Sheet molding compound A thermoset composite that can be formed into strong and stiff body components.

Shelf life The length of time the manufacturer recommends that a material should be kept before use.

Shim A thin metal piece used behind panels to bring them into alignment.

Shock tower The reinforced body areas for holding the upper parts of the suspension system.

Shop layout The arrangement of work areas, storage, aisles, office, and other spaces in a shop.

Shop manual A term often used for service manual.

Shop tool A major tool that the shop owner usually provides.

Short circuit An electrical leakage between two conductors or to ground.

Short method A partial replacement of adhesive when installing new fixed glass.

Show through Sand scratches in an undercoat that are visible through the topcoat.

Shrinking The act of removing a bulge from metal by hammering with a hammer and dolly, with or without heat.

Shrinking dolly A hand tool used with a shrinking hammer to reduce the surface area of metal without using heat.

Shrinking hammer A hand tool used with a shrinking dolly to reduce the surface area of metal without using heat.

Side sway Damage that occurs when an impact to the side of a vehicle causes the frame to bend or wrinkle.

Silicon carbide An abrasive used in sanding or grinding grit.

Silicone An ingredient in waxes and polishes which makes them feel smooth.

Silicone adhesive An adhesive used to repair torn vinyl trim and upholstery.

Silicone-treated graining paper Special paper that is used to create the final grain pattern in vinyl during upholstery repair.

Simulated seam tape Tape that is used to provide the appearance of seams in a spray-on vinyl roof covering.

Single coat Passing one time over the surface with each stroke overlapping the previous coat by 50 percent.

Single pull system A straightening system capable of only pulling in one direction at a time.

Siphon feed gun A type of paint spray gun in which the paint is drawn out of the container by vacuum action.

Skin The outer panel.

Skinning The formation of a thin tough film on the surface of a liquid paint film.

Slide caliper rule A rule used to measure inside or outside dimensions and the depth of a hole.

Slide hammer A hand-held tool having a hammer head that is slid along a rod and against a stop so that it pulls against the object to which the rod has been fastened.

Snap fit A joining technique in which the parts are forced over a lip or into an undercut retaining ring.

Sodium hydroxide powder A powder that is produced when the sodium azide pellets in an air bag are activated.

Soft-faced hammer A hammer having a head of plastic, wood, or other soft material.

Software Computer information stored in floppy disks, computer programs, and CDs.

Solder A mixture of tin, lead, antimony, or silver, that may be melted to fill dents and cracks in metal or to join wires in an electrical circuit.

Soldering A joining process in which the base metal is heated enough to allow the solder to melt and make an adhesive bond.

Solder paddle A wooden spatula-type tool used for applying and spreading body solder.

Solids The percentage of solid material in paint after solvents have evaporated.

Solvency The ability of a liquid to dissolve resin or any other material.

Solvent A liquid which will dissolve something, such as plastic.

Solvent cement A thin liquid that partly dissolves plastic materials so they may be bonded together.

Solvent popping The blisters that form on a paint film that are caused by trapped solvents.

Sound deadening material A pad or sheet of plastic material that absorbs sound.

Space frame A variation of unitized body construction in which molded plastic panels are bonded with adhesive or mechanical fasteners to a space frame.

Spanner socket A specialty tool used for special applications, such as the removal of antennas, mirrors, and radio trim nuts.

Specialty shops A shop that specializes in a certain area, such as frame straightening, wheel alignment, upholstering, or custom painting operations.

Specific gravity The ratio of the weight of a specific volume of a substance in the air compared to the weight of an equal volume of water.

Specifications Data supplied by the manufacturer covering all measurements and quantities of the vehicle.

Spectrophotometer An instrument to measure color.

Speed file A long holder used with strips of coarse sandpaper to smooth a work area.

Spider webbing A custom painting effect produced by forcing acrylic lacquer from the spray gun in the form of fibrous thread.

Spitting A paint spray gun problem caused by dried-out packing around the fluid needle valve.

Spontaneous combustion A process of material igniting by itself.

Spoon A tool used in the same manner as a dolly designed for use in confined areas and to pry panels back into position.

Spot cutter bit A tool used to cut through the welds on a panel.

Spot putty A plastic-like material used for filling small holes or sand scratches.

Spot repair A type of repair in which a section of a car smaller than a panel is repaired and refinished.

Spot weld A weld in which an arc is directed to penetrate both pieces of metal.

Spray gun A hand-held painting tool powered by air pressure that atomizes liquids, such as paint.

Spray mask A thin film that is sprayed on the surface to be decorated so a design is cut through the film and the desired portions may be removed before paint is applied.

Spray-on roof covering A vinyl coating sprayed in place.

Spray pattern A cross section of the spray.

Spreader A rigid rectangular piece of plastic used to spread body filler.

Spread ram A tool having two jaws that are forced apart when the tool is activated.

Squeegee A flexible rubber-like tool used to apply body putty or filler.

Stationary rack system A system that can be used to repair extensive collision damage.

Stationary section The permanent assemblies or components of a vehicle that cannot be moved.

Steel A ferrous metal used in the construction of a vehicle and as a substrate for paint, which must be painted to prevent corrosion.

Steering alignment specialist A technician who specializes in alignment and wheel balancing, as well as repairing steering mechanisms and suspension systems.

Steering axis inclination The inward tilt of the steering knuckle.

Steering system The mechanism that enables the driver to turn the wheels to change the direction of a vehicle's movement.

Stencil An impervious material into which designs have been cut.

Stitch welding The use of intermittent welds to join two or more parts.

Stools A low seat equipped with wheels.

Stop A check block or spacer used in movable glass installations.

Storage battery The device that converts chemical energy into electrical energy.

Strainer A fine mesh screen that is used to remove small lumps of dirt or other debris from a liquid.

Straight-in damage The damage that results from a direct impact.

Strength (1) The measure of the ability of a pigment to hide color. (2) The integrity of a structure.

Stress To relieve or take tension off a part.

Stress line The low area in a damaged panel that usually starts at the point of impact and travel outward.

Stretching The deformation of metal under tension.

Striker pin A bolt that can be adjusted laterally, vertically, fore, and aft to achieve door clearance and alignment.

Striker plate That portion of the door lock that is mounted on the body pillar.

Stringer bead A weld made by moving the electrode in a straight continuous line.

Striping brush A small brush used to apply stripes and other designs by hand.

Striping tool A tool with a small paint container and a brass wheel that applies paint as the tool is moved along the surface.

Stripper A term used for paint remover.

Stripping The act of removing paint by applying a chemical which softens and lifts it, by using air-powered blasting equipment, or by power sanding.

Structural adhesive A strong flexible thermosetting adhesive.

Structural integrity The body strength and ability of a vehicle to remain in one piece.

Structural member A load-bearing portion of the body structure that affects its over-the-road performance or crash worthiness.

Structural panel A panel used in a unibody that becomes a part of the whole unit and is vital to the strength of the body.

Stub frame A unibody with no center rail portions but with front and rear stub sections.

Strut suspension A suspension system that attaches to the spring tower and lower control arm.

Stud A headless bolt having threads on one or both ends.

Stud welding The joining of a metal stud or similar part to a work piece.

Subassembly An assembly of several parts that are put together before the whole is attached.

Sub frame A unitized body frame only having the front and rear stub sections of frame rails.

Sublet Repair or services sent out to another shop.

Submember A box- or channel-like reinforcement that is welded to a vehicle floor.

Substrate The surface that is to be worked on.

Suction cup A rubber or plastic cup-like device that is used to hold and position large sections of glass.

Sunroof A vehicle roof having a panel that slides back and forth on guide tracks.

Sunroof panel A transparent section of the roof that can be removed or slid into a recess for light or ventilation.

Support rod A metal rod that supports the head liner in a vehicle.

Surface preparation To prepare an old surface for refinishing or painting.

Surface rust Rust found on the outside of a panel that has not penetrated the steel.

Surfacer A heavily-pigmented paint that is applied to a substrate to smooth the surface for subsequent coats of paint.

Surface-scratching method The scratching of an arc welding electrode across the work to form the starting arc.

Surfacing mat A thin fiberglass mat used as the outer layer when making repairs.

Surform A surface forming grater file with open, rasp-like teeth.

Suspension system The springs and other components supporting the upper part of a vehicle on its axles and wheels.

Swirl remover A term used for polishing cream.

Symmetrical design A design in which both sides of a unibody are identical in structure and measurement.

Syntactic foam A resin and catalyst system that contains glass or plastic spheres used to fill rusted areas in door sills and rocker panels.

Tack The stickiness of a paint film or adhesive.

Tack cloth Cheesecloth that has been treated with nondrying varnish to make it tacky to pick up dust and lint.

Tack coat A light dusting coat that is allowed to become tacky before applying the next coat.

Tack rag A resin-impregnated cheesecloth used to pick up small dust and lint from a surface before being painted.

Tack weld A temporary weld to hold parts in place during final welding operations.

Tap A device used to cut internal threads.

Tape measure A retractable measuring tool.

Tapping technique The momentary touching of an arc welding electrode to the work as a means of striking an arc.

Tap and die The tools used to restore threads or to cut new threads.

Technical pen A pen used to draw pinstripes by hand on a vehicle.

Temperature-indicating crayon A temperature sensitive crayon used to mark across a weld area to monitor its temperature.

Temperature make-up system A heating/cooling system that filters and conditions the air before it enters a paint spray booth.

Tempered glass Glass that has been heat-treated.

Tensile strength The resistance to distortion.

Terminal A mechanical fastener attached to wire ends.

Test light An electrical test device that will light when voltage is present in a circuit.

Texturing agent The material that is added to paint to produce a bumpy texture.

Thermoplastic A plastic material that softens when heated and hardens when cooled.

Thermosetting A solid that will not soften when it is heated.

Thermosetting plastic A resin that does not melt when it is heated.

Thinner A solvent used to thin lacquers and acrylics.

Three-stage Three paint layers that produce a pearlescent appearance consisting of a basecoat, a midcoat, and a clearcoat.

Thrust line An imaginary line parallel with the rear wheels.

TIG An abbreviation for tungsten inert gas arc welding.

Tinning The melting of a solder and flux onto an area to be soldered.

Tint (1) A light color, usually pastel. (2) To add color to white or another color.

Tinted glass Window glass having a color tint.

Tinting strength The ability of a pigment to change the color of a paint to which it is added.

Tint tone The shade that results when a color is mixed with white paint.

Toe The position of the front of a wheel when compared to the rear of the wheel.

Toe-in A condition whereby the front edge of the wheels are closer together than the rear edge of the wheels.

Toe-out A condition whereby the rear edge of the wheels are closer together than the front edge of the wheels.

Tolerance The acceptable variation, plus or minus, of vehicle dimensions as provided by the manufacturer.

Topcoat The top layer of paint applied to a substrate.

Torque box A structural component provided to permit some twisting as a means of absorbing road and collision impact shock.

Torsion bar A metal bar that is twisted as a lid is closed to provide a spiral tension.

Total loss A situation whereby the cost of repairs would exceed the vehicle value.

Touch-up gun A paint spray gun, similar to a conventional spray gun, but with a smaller capacity used for touch-up, stenciling, and small detail.

Touch-up paint A small container of paint matched to the factory color used to fill small chips in a vehicle finish.

Tower The upright portion of a frame-straightening systems.

Tower cut The same as a nose, but includes the shock tower or strut tower.

Toxic fumes Harmful fumes that can cause illness or death.

Toxicity The biological property of a material reflecting its inherent capacity to produce injury or an adverse effect due to overexposure.

Tracer A colored coding stripe on an electrical wire for identification when tracing a circuit.

Tracking The ability of the rear wheels of a vehicle to follow the front wheels.

Tracking gauge A tool used to detect and measure the misalignment of front and rear wheels.

Track molding A trim piece consisting of a metal track and a plastic insert.

Tram gauge An instrument used to check alignment and dimensions against factory specifications.

Transmitter/receiver lock system A system whereby a signal is transmitted from a small key ring transmitter to the receiver located in the door to activate the locking mechanism.

Transverse An engine positioned so that its crankshaft is parallel to the vehicle's axles.

Trim A decorative metal piece on a vehicle body.

Trim cement An adhesive used to attach upholstery and selected trim.

Tunnel A formation in the floor panel for transmission and drive shaft clearance on a rear-wheel-drive vehicle.

Turning radius The amount one front wheel turns sharper than the other.

Turpentine A solvent derived from the distillate of pine trees.

Twist Collision damage that causes distortion of the frame cross members.

Two component epoxy primer A primer material having two components that react after being mixed together.

Two-part A product supplied in two parts which must be mixed together in correct proportions immediately before use.

Two-stage Two coats of paint such as a basecoat and a clearcoat.

Two-tone Two different colors on a single paint scheme.

Ultrasonic plastic welding A technique of repairing rigid auto plastics in which welding time is controlled by a power supply.

Ultrasonic stud welding The variation of a shear joint used to join plastic parts whereby a weld is made along a stud's circumference.

Ultraviolet (UV) light That part of the invisible light spectrum which is responsible for degradation of paints.

Ultraviolet (UV) stabilizer A chemical added to paint to absorb ultraviolet radiation.

Underbody The lower portion of a vehicle that contains the floor pan, trunk floor, and structural reinforcements.

Undercoat The first coat of a primer, sealer, or surfacer.

Undercoating (1) The second coat of a three-coat finish; the first coat in repainting. (2) The coating or sealer on the underside of panels to help prevent rust and deaden sound.

Undercut A groove in the base metal adjacent to a weld and left unfilled by weld metal.

Unibody A vehicle style whereby parts of the body structure serve as support for overall vehicle strength.

Unibody construction A vehicle construction type in which the body and underbody form an integral structural unit.

Universal measuring system A measuring system having frame-mounted devices that can be adjusted for various vehicle bodies.

Universal thinner A solvent that is used to thin lacquers and to reduce enamels.

Upholsterer One whose expertise is the repair or replacement of interior surface materials.

Upholstery ring pliers Pliers that are used to remove or install upholstery rings.

Upsetting The deformation of metal under compression.

Up-stop A component that limits the upward travel of the lift channel.

Urethane A type of paint or polymer coating noted for its toughness and abrasion resistance.

Utility knife A hand-held cutting tool having a replaceable retractable blade.

UV stabilizer A chemical added to paint to absorb ultraviolet radiation.

Vacuum Any pressure below atmospheric pressure.

Vacuum cleaner A portable suction device used to clean vehicle interiors.

Vacuum cup puller A large suction cup used as a dent puller.

Vacuum patch A device that is placed over a glass repair area to withdraw all air to ensure that all voids are filled with resin.

Value (1) The darkness or lightness of a color. (2) The fair cost or price of an item.

Vapor A state of matter; a gaseous state.

Vehicle (1) A car or a truck. (2) All of a paint except the pigment including solvents, diluents, resins, gums, and dryers.

Vehicle identification number (VIN) A number that is assigned to vehicles by the manufacturer for registration and identification purposes.

Veil A term used for a surfacing mat.

Veiling The formation of a web or strings in a paint as it emerges from a spray gun.

Ventilation fan An electrical device used to remove vapors and fine particles from the work area.

Vibrating knife A knife with a rapidly moving blade that cuts through polyurethane adhesives.

Vinyl A class of material which can be combined to form vinyl polymers used to make chemical resistant finishes and tough plastic articles.

Vinyl-coated fabrics Any material with a plastic protective or decorative layer bonded to a fabric base that provides strength.

Vinyl paint Any paint material applied to a vinyl top to restore color.

Viscosity (1) The consistency or body of a paint. (2) The thickness or thinness of a liquid.

Visual estimate A guess by an experienced estimator relative to the cost of repairing damage.

VOC An abbreviation for volatile organic compound.

Volatile A material that vaporizes easily.

Volatile organic compound A hydrocarbon that readily evaporates into the air and is extremely flammable.

Volatility The tendency of a liquid to evaporate.

Voltage The electrical pressure that causes current to flow.

Voltmeter An electrical device used to measure voltage.

Volt A unit of measure of electrical pressure.

Wage The amount of money paid to workers, generally computed on a "per hour" basis.

Warpage The distortion of a panel during heat shrinking.

Wash thinner A low cost solvent used to clean spray guns and other equipment.

Water/air shield A deflector built into a vehicle door.

Waterproof sandpaper Sandpaper that may be used with water for wet sanding.

Water spotting A condition created when water evaporates on a finish before it is thoroughly dry.

Wax (1) A slippery solid sometimes added to paints to add some property. (2) A prepared material used to shine or improve a surface.

Weathering A change in paint film caused by natural forces such as sunlight, rain, dust, and wind.

Weather stripping A rubber-like gasket used to keep dirt, air, and moisture out of the passenger or trunk compartments of a vehicle.

Weave bead A wide weld bead made by moving the electrode back and forth in a weaving motion.

Weld The act of joining two metal or plastic pieces together by bringing them to their melting points.

Welding The joining method that involves melting and fusing two pieces of material together to form a permanent joint.

Weld through primer A primer applied to a joint before it is welded to help prevent galvanic corrosion.

Wetcoat A heavy coat of paint.

Wet sanding Sanding using a water resistant, ultra-fine sandpaper and water to level paint.

Wet spot A discoloration caused where paint fails to dry and adhere uniformly.

Wet-on-wet finish A technique of applying a fresh coat of paint over an earlier coat which has been allowed to flash but not cure.

Wheel alignment The positioning of suspension and steering components to assure a vehicle's proper handling and maximum tire wear.

Wheel balancing The act of properly distributing weight around a tire and wheel assembly to maintain a true running wheel perpendicular to its rotating axis.

Wheelbase The distance between front and rear axles.

Wheelhouse The deep curved panels that form compartments in which the wheels rotate.

Whipping The improper movement of a paint spray gun that wastes energy and material.

Wind cord A rope-like trim placed around doors to help seal and decorate the openings.

Wind lace A rope-like trim placed around doors to help seal and decorate the openings.

Window channel The grooves, guides, or slots in which the glass slides up and down or back and forth.

Window regulator The door-mounted mechanism that provides a means of cranking the window up and down.

Window stop A device inside a door used to limit glass height and depth.

Window tool Tools required to properly remove and install window glass.

Windshield pillar The structural member that attaches the body to the roof panel.

Wiring diagram Drawings that show where wires are routed and how components are arranged.

Wiring harness Electrical wires gathered in a bundle.

Wood grain transfer A plastic transfer sheet used to simulate wood on the sides of a vehicle.

Work hardening Metal that has become stiffer and harder in the stretched areas due to permanent stresses.

Wrench A hand tool used to turn fasteners.

Wrinkle (1) The pattern formed on the surface of a paint film by improperly formulated or specially formulated coatings. (2) The appearance of tiny ridges or folds in paint film.

Wrinkling A surface distortion that occurs in a thick coat of enamel before the underlayer has properly dried.

X-checking The process of taking measurements and comparing them to corresponding dimensions on the opposite side of the vehicle to reveal damage.

X-frame A frame design that does not rely on the floor pan for torsional rigidity.

Zahn cup A paint cup with a hole in the bottom that is used to measure a material's viscosity.

Zebra effect A streaky looking metallic finish, usually caused by uneven application.

Zero plane A plane that divides the datum plane into front, middle, and rear sections.

Zinc A metal coating used to prevent corrosion.

Zinc te A primer for aluminum and other nonferrous metals.

Zoning A method of systematically observing a damaged vehicle.

Zoning ordinance The law that limits the type of businesses allowed in a particular area or zone.